Mathematics and Digital Signal Processing

Mathematics and Digital Signal Processing

Editor

Pavel Lyakhov

MDPI • Basel • Beijing • Wuhan • Barcelona • Belgrade • Manchester • Tokyo • Cluj • Tianjin

Editor
Pavel Lyakhov
Mathematical Modeling
North-Caucasus Federal
University
Stavropol
Russia

Editorial Office
MDPI
St. Alban-Anlage 66
4052 Basel, Switzerland

This is a reprint of articles from the Special Issue published online in the open access journal *Applied Sciences* (ISSN 2076-3417) (available at: www.mdpi.com/journal/applsci/special_issues/ Mathematics_Digital_Signal_Processing).

For citation purposes, cite each article independently as indicated on the article page online and as indicated below:

LastName, A.A.; LastName, B.B.; LastName, C.C. Article Title. *Journal Name* **Year**, *Volume Number*, Page Range.

ISBN 978-3-0365-1476-5 (Hbk)
ISBN 978-3-0365-1475-8 (PDF)

© 2021 by the authors. Articles in this book are Open Access and distributed under the Creative Commons Attribution (CC BY) license, which allows users to download, copy and build upon published articles, as long as the author and publisher are properly credited, which ensures maximum dissemination and a wider impact of our publications.

The book as a whole is distributed by MDPI under the terms and conditions of the Creative Commons license CC BY-NC-ND.

Contents

About the Editor vii

Nikolay Chervyakov, Pavel Lyakhov and Nikolay Nagornov
Analysis of the Quantization Noise in Discrete Wavelet Transform Filters for 3D Medical Imaging
Reprinted from: *Applied Sciences* **2020**, *10*, 1223, doi:10.3390/app10041223 1

Junseok Lim
Maximum Correntropy Criterion Based l_1-Iterative Wiener Filter for Sparse Channel Estimation Robust to Impulsive Noise
Reprinted from: *Applied Sciences* **2020**, *10*, 743, doi:10.3390/app10030743 29

Tatiana Klishkovskaia, Andrey Aksenov, Aleksandr Sinitca, Anna Zamansky, Oleg A. Markelov and Dmitry Kaplun
Development of Classification Algorithms for the Detection of Postures Using Non-Marker-Based Motion Capture Systems
Reprinted from: *Applied Sciences* **2020**, *10*, 4028, doi:10.3390/app10114028 37

Dmitry Kaplun, Mikhail Golovin, Alisa Sufelfa, Oskar Sachenkov, Konstantin Shcherbina, Vladimir Yankovskiy, Eugeniy Skrebenkov, Oleg A. Markelov and Mikhail I. Bogachev
Three-Dimensional (3D) Model-Based Lower Limb Stump Automatic Orientation
Reprinted from: *Applied Sciences* **2020**, *10*, 3253, doi:10.3390/app10093253 53

Dmitry Kaplun, Sergey Aryashev, Alexander Veligosha, Elena Doynikova, Pavel Lyakhov and Denis Butusov
Improving Calculation Accuracy of Digital Filters Based on Finite Field Algebra
Reprinted from: *Applied Sciences* **2019**, *10*, 45, doi:10.3390/app10010045 67

Ali Dehghan Firoozabadi, Pablo Irarrazaval, Pablo Adasme, David Zabala-Blanco, Hugo Durney, Miguel Sanhueza, Pablo Palacios-Játiva and Cesar Azurdia-Meza
Multiresolution Speech Enhancement Based on Proposed Circular Nested Microphone Array in Combination with Sub-Band Affine Projection Algorithm
Reprinted from: *Applied Sciences* **2020**, *10*, 3955, doi:10.3390/app10113955 81

Dmitry Kaplun, Alexander Voznesensky, Sergei Romanov, Valery Andreev and Denis Butusov
Classification of Hydroacoustic Signals Based on Harmonic Wavelets and a Deep Learning Artificial Intelligence System
Reprinted from: *Applied Sciences* **2020**, *10*, 3097, doi:10.3390/app10093097 111

Nikita S. Pyko, Svetlana A. Pyko, Oleg A. Markelov, Oleg V. Mamontov and Mikhail I. Bogachev
Quantification of the Feedback Regulation by Digital Signal Analysis Methods: Application to Blood Pressure Control Efficacy
Reprinted from: *Applied Sciences* **2019**, *10*, 209, doi:10.3390/app10010209 125

Yutu Yang, Xiaolin Zhou, Ying Liu, Zhongkang Hu and Fenglong Ding
Wood Defect Detection Based on Depth Extreme Learning Machine
Reprinted from: *Applied Sciences* **2020**, *10*, 7488, doi:10.3390/app10217488 141

Nikolay Chervyakov, Pavel Lyakhov, Mikhail Babenko, Irina Lavrinenko, Maxim Deryabin, Anton Lavrinenko, Anton Nazarov, Maria Valueva, Alexander Voznesensky and Dmitry Kaplun
A Division Algorithm in a Redundant Residue Number System Using Fractions
Reprinted from: *Applied Sciences* **2020**, *10*, 695, doi:10.3390/app10020695 . . . . **155**

About the Editor

Pavel Lyakhov

Pavel Lyakhov is currently the Head of the Department of Mathematical Modeling, North-Caucasus Federal University. He graduated in mathematics from Stavropol State University, in 2009, where he also received a Ph.D. degree in mathematics, in 2012. He has been working with North-Caucasus Federal University, since 2012. Currently leads research projects: RFBR grant 19-07-00130-A -effective tools for intellectual analysis of visual information based on convolutional neural networksand Russian Federation President grant MK-3918.2021.1.6 -performance digital medical imaging circuits based on parallel mathematics. His research interests include high-performance computing, residue number systems, digital signal processing, image processing, and medical imaging.

Article

Analysis of the Quantization Noise in Discrete Wavelet Transform Filters for 3D Medical Imaging

Nikolay Chervyakov, Pavel Lyakhov and Nikolay Nagornov *

Department of Applied Mathematics and Mathematical Modeling, North-Caucasus Federal University, Stavropol 355017, Russia; k-fmf-primath@stavsu.ru (N.C.); ljahov@mail.ru (P.L.)
* Correspondence: sparta1392@mail.ru; Tel.: +7-962-451-3247

Received: 14 January 2020; Accepted: 8 February 2020; Published: 11 February 2020

Abstract: Denoising and compression of 2D and 3D images are important problems in modern medical imaging systems. Discrete wavelet transform (DWT) is used to solve them in practice. We analyze the quantization noise effect in coefficients of DWT filters for 3D medical imaging in this paper. The method for wavelet filters coefficients quantizing is proposed, which allows minimizing resources in hardware implementation by simplifying rounding operations. We develop the method for estimating the maximum error of 3D grayscale and color images DWT with various bits per color (BPC). The dependence of the peak signal-to-noise ratio (PSNR) of the images processing result on wavelet used, the effective bit-width of filters coefficients and BPC is revealed. We derive formulas for determining the minimum bit-width of wavelet filters coefficients that provide a high (PSNR ≥ 40 dB for images with 8 BPC, for example) and maximum (PSNR = ∞ dB) quality of 3D medical imaging by DWT depending on wavelet used. The experiments of 3D tomographic images processing confirmed the accuracy of theoretical analysis. All data are presented in the fixed-point format in the proposed method of 3D medical images DWT. It is making possible efficient, from the point of view of hardware and time resources, the implementation for image denoising and compression on modern devices such as field-programmable gate arrays and application-specific integrated circuits.

Keywords: discrete wavelet transform; medical imaging; 3D image processing; quantization noise

1. Introduction

Medical imaging uses many different methods such as magnetic resonance (MR) imaging [1–8], radiography [4,9–11], radionuclide [8,12], optical [11,13,14], ultrasound [1,15] and medical robotics [16,17]. The typical medical imaging system consists of three components (Figure 1): data acquisition, data consolidation and data processing. The data acquisition card, which filters incoming data, is the most cost-sensitive system card. Usually, a diagnostic imaging system will consist of multiple data acquisition cards. Once the data is compensated and filtered in scanners, it is sent to the data consolidation card for buffering and data alignment. Once the data has been collected, it is sent to the image processing cards [18]. These cards perform heavy-duty filtering and the most algorithm-intensive image reconstruction. Modern field-programmable gate array (FPGA) devices are widely used in data consolidation, and image processing for sophisticated application algorithms implementation including pattern recognition, image enhancement and data compression [19,20].

Denoising of 2D and 3D medical images is an important problem in modern medical imaging systems. The noisy pattern is not always bad in medical images, but in most cases is a problem. MR images are inherently noisy and thus filtering methods are required to improve the data quality [5]. Rheological methods of increasing MR elastography resolution determine viscoelastic properties through wave inversion, which is highly ill posed and sensitive to noise [1]. In radiology using computed tomography (CT) or related morphological imaging modalities, noise affects the analysis of anatomical structures and thus impedes diagnostic applications [11]. Low dose radiation exposure for patient safety leads to noisy and low-contrast fluoroscopic sequences [11]. The reconstruction process of the positron emission tomography images includes inherent multiplicative noise, which prevents the analysis of visual data [12]. In optical CT for retinal imaging as another example use case, noise limits the measurement of structural features in the human eye, e.g., retinal layer properties [11]. Denoising facilitates visual data interpretation from echocardiography [15].

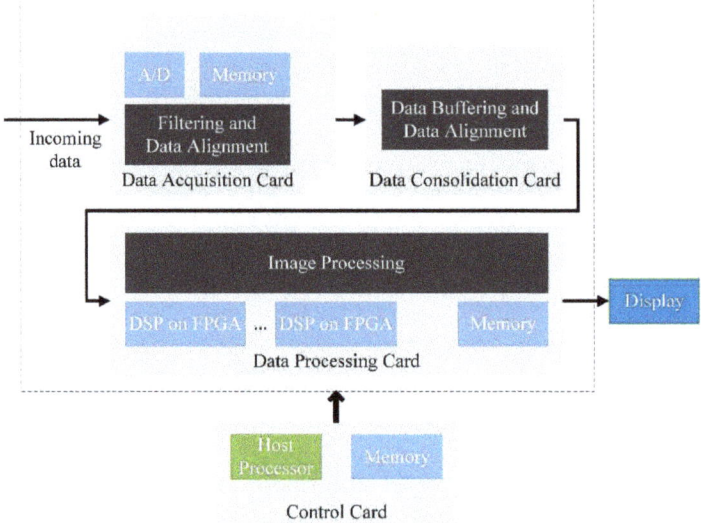

Figure 1. The typical medical imaging system.

Medical imaging systems produce increasingly accurate images with improved quality using higher spatial resolutions and bit-depths with advances in scanning technology and digital devices. Such improvements increase the amount of information that needs to be processed, transmitted and stored. This is especially true when using 3D scanning technology [4]. For example, four sets of positron emission tomography medical images of one patient may require more than 4 GB of storage space [21]. Video recording of a relatively short retinal peeling procedure may require over 40 GB of memory storage [14]. The capacity of hard drives is on average 1–2 TB with the current level of storage technology development. Thus, the compression of 3D medical images is also an important problem in modern medical imaging systems.

Various transforms are used to solve problems of 2D and 3D medical images denoising and compression in practice. The most common of them are discrete Fourier transform (DFT) [3,7,14,22] and discrete wavelet transform (DWT) [1,9,11,14]. DFT is widely used in the frequency domain but the domain characteristics disappeared after it. We cannot determine the time position and the degree of intensity after signal DFT. It is not possible to describe the local properties of the time domain of the image. DWT solves these problems because it allows obtaining both frequency and time information about a signal [23,24]. 2D and 3D images DWT is performed by convolution with a pair of lowpass and highpass wavelet filters of filter bank that highlight main and detailed information respectively. Denoising and compression of images are performed by detailed information manipulating in modern algorithms such as set partitioning in hierarchical trees (SPIHTs) [25] and embedded zerotrees of wavelet transforms (EZWs) [26]. The convolution operation has high computational complexity. Hardware implementation on modern microelectronic devices such as field-programmable gate arrays (FPGAs) and application-specific integrated circuits (ASICs) working with fixed-point numbers is one of the ways to improve its characteristics [27–29]. Quantization noise occurs when converting wavelet filters coefficients into this format, due to which convolution is performed with an error. The question arises about the accuracy of wavelet filters coefficients representation in the device's memory, which is efficient in terms of resources and enough to achieve the required quality of image processing. A novel area-efficient high-throughput 3D DWT architecture for real-time medical imaging based on distributed arithmetic is proposed by the authors [30]. The design and implementation of 3D Haar wavelet transform with transpose based computation and dynamic partial reconfiguration for 3D medical image compression are presented in [31]. The implementation of positron emission tomography using DWT on FPGA is proposed by authors [32]. In paper [33] described the architecture based on the use of DWT for biomedical signals compression. The design and implementation of context-based adaptive variable length coding and comparative analysis of trade-off offered by DWT for 3D medical image compression systems are described by authors [34]. In [35] presented the design and implementation of 3D DWT with a transpose-based method for medical image compression on FPGA. Experimental results from [36] showed that the system constructed a 1D DWT system based on FPGA can filter the noise and extract the electroencephalogram (EEG) signal well. The design and implementation on FPGA of 3D DWT using Daubechies wavelets with a transpose-based method for medical image compression are presented in [37]. The design and implementation of distributed arithmetic architectures of 3D DWT with a hybrid method for medical image compression are presented in [38]. Authors [39] presented the FPGA-based embedded system design using DWT and its evaluation for a pre-processing stage of EEG signal analysis. A detailed review of FPGA and ASIC architectures for DWT implementation in biomedical and intelligent applications, which can be designed either for higher-accuracy or for low-power consumption is provided by the authors [29]. In [40], the authors showed that DWT along with Gaussian filtering shows better results in removing the noise and smoothes the electrocardiogram signals. Authors [41] described the design and implement a complete hardware model based on DWT for EEG data compression and reconstruction on FPGA. A framework is offered in [42] based on DWT using linear and non-linear classifiers for detecting an epileptic seizure from EEG data recorded from normal subjects and an epileptic patient. There are no references to selected bit-width of wavelet filters coefficients in the materials studied about the hardware implementation of medical images DWT on FPGA and ASIC [29–32,34–42]. Authors [33] quantized wavelet filters coefficients by 16 bits, but there is no rationale for this choice. The problem of analyzing the quantization noise effect in wavelet filters coefficients for 2D grayscale and color images DWT with 8 bits per color (BPC) was solved in [43].

Analysis of 3D medical images DWT result quality dependence on noise arising from filters coefficients quantizing of wavelet with compact support is the purpose of this work. Particular attention is paid to determining the minimum bit-width of wavelet filters coefficients, at which this noise does not have a significant impact on the 3D medical images DWT result ($PSNR \geq 40$ dB for images with 8 BPC, for example), or does not affect it at all ($PSNR = \infty$). The values $PSNR \geq 40$ dB describes the difference between the two images with 8 BPC almost imperceptible for human eyes [44,45]. The value $PSNR = \infty$ for identical images.

2. Materials and Methods

DWT is a signal transform using a filter bank, which is a convolution of the input data with wavelet filters that translate them from a time representation into a time-frequency domain. Wavelet filters F of filter bank consist of coefficients $f_{F,i}$, where $i = 1, \ldots, k$ and k is the number of coefficients. Coefficients of lowpass and highpass wavelet filters of decomposition (LD, HD) and reconstruction (LR, HR) are related by equation [27]

$$f_{HD,i} = (-1)^{i+1} f_{LD,k-1-i}, \quad f_{LR,i} = f_{LD,k-1-i}, \quad f_{HR,i} = (-1)^i f_{LD,i}. \tag{1}$$

We shall consider only wavelets with compact support [46]. Daubechies wavelets $db(k/2)$ (where $db1$ with $k = 2$ is Haar wavelet), symlets $sym(k/2)$ and coiflets $coif(k/6)$ are the most common ones.

Consider a 3D digital medical image I of X rows, Y columns and Z frames as a function $I(x, y, z)$, where $0 \leq x \leq X-1, 0 \leq y \leq Y-1$ and $0 \leq z \leq Z-1$ are the spatial coordinates of I. Thus, voxel values (analogues of 2D pixels for 3D space) are represented as $I(x, y, z)$ for grayscale images and as $I(x, y, z, c)$ for color images, where c is the color number (for example, $c = 1, 2, 3$—red, green and blue colors respectively for RGB images). We assumed that all image voxels are isotropic [47], hereinafter.

Convolution of a 3D image with wavelet filters is performed by formulas

$$I'(x,y,z) = \sum_{i=1}^{k} I(x-i,y,z) \cdot f_{F,i},\ I''(x,y,z) = \sum_{i=1}^{k} I'(x,y-i,z) \cdot f_{F,i},$$

$$I'''(x,y,z) = \sum_{i=1}^{k} I''(x,y,z-i) \cdot f_{F,i},$$

where I', I'' and I''' is the convolution results by strings, columns and frames respectively. 3D image DWT is performed by sequential convolution with wavelet filters (Figure 2) in the steps below.

1. Row analysis is performed by decomposing the original image I by rows with lowpass LD and highpass HD wavelet filters and downsampling, indicated by the symbol $\downarrow 2$ (for example, the array $[2, 8, 5, 1, -1, 3]$ is transformed into an array $[2, 5, -1]$ after the operation $\downarrow 2$).
2. Column analysis is performed by columns similar to the row analysis for coefficients obtained at stage 1.
3. Frame analysis is performed by frames similar to the row analysis for coefficients obtained at stage 2.

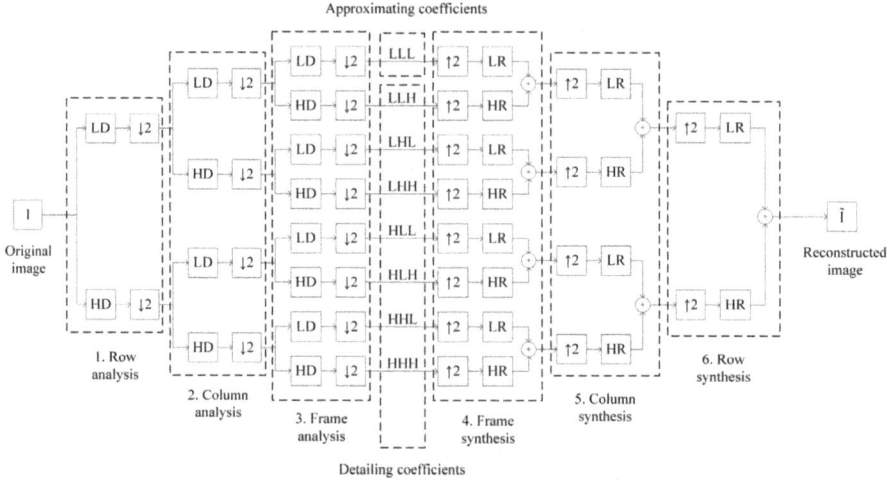

Figure 2. The scheme of 3D image discrete wavelet transform (DWT).

We get 8 sets of coefficients, $LLL, LLH, LHL, LHH, HLL, HLH, HHL$ and HHH, of image decomposition as a result of original image I analysis. These sets can be divided into approximating (LLL) and detailing ($LLH, LHL, LHH, HLL, HLH, HHL$ and HHH). Approximating coefficients correspond to the lowpass part of the signal and contain main information about the image I. Detailing coefficients to correspond to the highpass part of the signal and contain detailed information about the image I. 3D image denoising and compression are carried out by manipulating detailing coefficients ($LLH, LHL, LHH, HLL, HLH, HHL$ and HHH) of image decomposition.

4. Frame synthesis is performed by upsampling, indicated by the symbol ↑ 2 (for example, the array $[2, 5, -1]$ is transformed into an array $[2, 0, 5, 0, -1, 0]$ after the operation ↑ 2), of image decomposition coefficients by frames, reconstructing with lowpass *LR* and highpass *HR* wavelet filters and summation of the corresponding results.
5. Column synthesis is performed by columns similar to the frame synthesis for coefficients obtained at Stage 4.
6. Row synthesis is performed by rows similar to the frame synthesis for coefficients obtained at Stage 5.

We get the reconstructed image $\widetilde{I}$ as a result of image decomposition coefficients synthesis. Theoretically, the original image should be fully reconstructed since the scheme in Figure 2 has the perfect reconstruction property [48]. However, quantization noise occurs due to the digital format of wavelet filters coefficients representation in practice. Quantization noise distorts all image decomposition coefficients *LLL*, *LLH*, *LHL*, *LHH*, *HLL*, *HLH*, *HHL* and *HHH* as well as reconstructed image $\widetilde{I}$. The images DWT result may have a quality unacceptable for the task depending on the magnitude of quantization noise.

The question arises about the minimum bit-width of wavelet filters coefficients $f_{F,i}$, necessary for efficient software and hardware implementation of 3D images DWT on modern devices and enough for high-quality images processing. The speed of operations with a fixed-point number is higher than with a floating-point number on modern devices. This can be used to develop 3D medical imaging devices. Therefore, wavelet filters coefficients are quantized and converted into a fixed-point format in the proposed method by scaling by 2^n and rounding up

$$f_{F,i}^* = \lceil 2^n f_{F,i} \rceil. \tag{2}$$

Bit-width r of quantized wavelet filters coefficients $f_{F,i}^*$ can be determined by the formula $r = n + 1$ in this case. The digital image I^* processed according to the scheme in Figure 2 using quantized wavelet filters coefficients $f_{F,i}^*$. Voxel values of an image I^* should be normalized by scaling by 2^{-6n} (2^{-n} for each convolution, according to the scheme from Figure 2) and rounding down

$$\widetilde{I} = \lfloor 2^{-6n} I^* \rfloor. \tag{3}$$

We get only integers as a result of images DWT with unquantized coefficients. The quantization error of the wavelet filters coefficients rounded up is strictly redundant. Rounding down of the DWT results minimizes this error and cannot cause an error by itself. Rounding up and down operations are performed by discarding the fractional part of the number with the addition of one in the case of rounding up an integer. The rounding errors will have different signs and partially compensate each other for rounding in different directions. Rounding operations in this order require fewer resources for hardware implementation than rounding operations to the nearest integer. This is due to the fact wavelet filter coefficients are known a priori and their quantization with rounding up can be made in advance. Thus, wavelet filters coefficients will be used in the form of constants in the software and hardware part. The convolution is performed using arithmetic logic devices, and its result is rounded down by simply discarding the fractional part and does not require additional hardware and time costs.

We used the peak signal-to-noise ratio (*PSNR*) between two images (original image *I* and processed image $\widetilde{I}$) to quantify the image processing quality. The *PSNR* logarithmic nature makes it possible to clearly interpret results that differ slightly from each other. Other metrics usually only show a big difference. This characteristic is measured in decibels (dB) and is calculated by the following formula [49]

$$PSNR = 10 \log_{10} \left(\frac{(2^B - 1)^2}{MSE} \right) = 10 \log_{10} \left(\frac{M^2}{MSE} \right),$$

where: B is the image BPC; M is the maximum brightness of the image voxels (for example, $B = 8$ and $M = 2^8 - 1 = 255$ for 8-bit grayscale image and 24-bit RGB image); MSE is the mean square error of brightness, which is calculated for grayscale ($MSE_{grayscale}$) [50] and color (MSE_{color}) [51] 3D images by formulas

$$MSE_{grayscale} = \sum_{x=0}^{X-1}\sum_{y=0}^{Y-1}\sum_{z=0}^{Z-1} \frac{\left(I(x,y,z) - \widetilde{I}(x,y,z)\right)^2}{X \cdot Y \cdot Z},$$

$$MSE_{color} = \frac{1}{C}\sum_{c=1}^{C}\sum_{x=0}^{X-1}\sum_{y=0}^{Y-1}\sum_{z=0}^{Z-1} \frac{\left(I(x,y,z,c) - \widetilde{I}(x,y,z,c)\right)^2}{X \cdot Y \cdot Z}.$$

The value $PSNR = \infty$ for identical images. The image processing quality is considered high if $PSNR \geq Q$, where Q describes the difference between the two images almost imperceptible for human eyes. $Q = 40$ dB for images with 8 BPC [44,45]. We propose to generalize Q to the case of images with 12 and 16 BPC using formula

$$Q = 5B. \qquad (4)$$

Thus, Q is equal 40 dB, 60 dB and 80 dB for images with 8, 12 and 16 BPC respectively.

3. Results

3.1. Theoretical Analysis of the Maximum Error of the 3D Medical Images DWT

The error of 3D medical images DWT occurs as a result of wavelet filters coefficients conversion (quantization noise) by Formula (2). Convolutions, upsampling and the summing of convolution results cause an increase in this error. Rounding down normalized voxel values of the restored image also has an effect. Note the important facts.

1. The analyzing and synthesizing wavelet filters consist of the same coefficients, according to Formula (1), hence, the limited absolute errors of computations will also be equal. Therefore, within the framework of theoretical calculations, wavelet filters are classified only into lowpass L and highpass H ones.
2. The sums of the lowpass and highpass wavelet filter coefficients are equal $\sum_{i=1}^{k} f_{L,i} = \sqrt{2}$ and $\sum_{i=1}^{k} f_{L,i} = \sqrt{2}$, respectively [27].

We introduce the following notation.

1. $E_{j,F}$—limited absolute error (LAE) of calculating the value of the coefficient at the j-th stage, resulting from convolution with a sequence of wavelet filters F.
2. S_F—the exact value of the sum of the coefficients of the wavelet filter F.
3. $T_{j,F}$—the exact value of the calculations in the j-th stage, after convolution with a sequence of wavelet filters F.

The errors a of all image decomposition coefficients $LLL, LLH, LHL, LHH, HLL, HLH, HHL$ and HHH are separated into two groups a_ε ($\varepsilon = 1,2$) as a result of upsampling $\uparrow 2$. Figure 3 shows an example of the errors separation a_ε ($\varepsilon = 1,2,3,4$) at the upsampling by frames and columns, where $Y^* = (Y+k)/2 - 1$ and $Z^* = (Z+k)/2 - 1$. This situation is similar for upsampling by strings. Upsampling is applied three times during image restoration. We got eight groups of errors a_ε ($\varepsilon = 1,2,3,4,5,6,7,8$) as a result. Thus, we would add an additional index ε to the introduced notations, which denotes calculations by the spatial characteristics of wavelet filters coefficients.

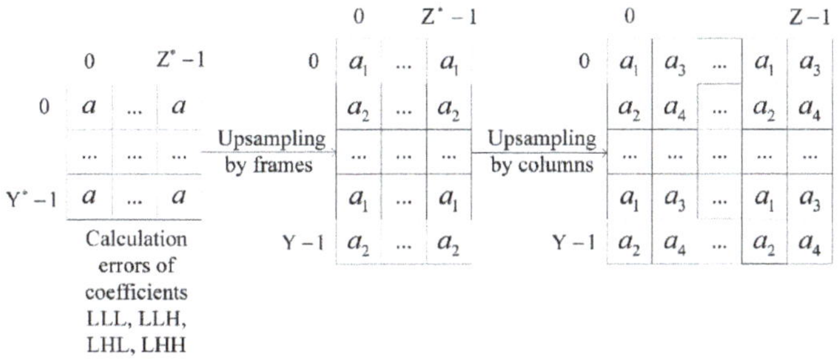

Figure 3. The scheme of the errors separation with upsampling by frames and columns.

Next, we carried out analysis calculations for an estimation of the maximum error of the 3D medical images DWT.

Stage 1. Wavelet filters coefficients quantization. Let us calculate the exact values of the coefficients sums S_F, $S_{F,\varepsilon}$ and errors $E_{1,F}$, $E_{1,F,\varepsilon}$ of rounding up filters L and H scaled coefficients.

$$S_L = \sum_{i=1}^{k} 2^n f_{L,i} = 2^n \sum_{i=1}^{k} f_{L,i} = 2^n \cdot \sqrt{2} = 2^{n+\frac{1}{2}}, S_H = \sum_{i=1}^{k} 2^n f_{H,i} = 2^n \sum_{i=1}^{k} f_{H,i} = 2^n \cdot 0 = 0,$$

$$S_{L,1} = \sum_{i=1}^{\frac{k}{2}} 2^n f_{L,2(i-1)}, S_{L,2} = \sum_{i=1}^{\frac{k}{2}} 2^n f_{L,2i-1}, S_{H,1} = \sum_{i=1}^{\frac{k}{2}} 2^n f_{H,2(i-1)}, S_{H,2} = \sum_{i=1}^{\frac{k}{2}} 2^n f_{H,2i-1},$$

$$E_{1,L} = \sum_{i=1}^{k} \left(\lceil 2^n f_{L,i} \rceil - 2^n f_{L,i} \right), E_{1,H} = \sum_{i=1}^{k} \left(\lceil 2^n f_{H,i} \rceil - 2^n f_{H,i} \right),$$

$$E_{1,L,1} = \sum_{i=1}^{\frac{k}{2}} \left(\lceil 2^n f_{L,2(i-1)} \rceil - 2^n f_{L,2(i-1)} \right), E_{1,L,2} = \sum_{i=1}^{\frac{k}{2}} \left(\lceil 2^n f_{L,2i-1} \rceil - 2^n f_{L,2i-1} \right),$$

$$E_{1,H,1} = \sum_{i=1}^{\frac{k}{2}} \left(\lceil 2^n f_{H,2(i-1)} \rceil - 2^n f_{H,2(i-1)} \right), E_{1,H,2} = \sum_{i=1}^{\frac{k}{2}} \left(\lceil 2^n f_{H,2i-1} \rceil - 2^n f_{H,2i-1} \right).$$

Stage 2. Row decomposition. Let us calculate the exact values $T_{2,F}$ and errors $E_{2,F}$ of row decomposition with filters L and H.

$$T_{2,L} = S_L \cdot M, E_{2,L} = E_{1,L} \cdot M, E_{2,H} = E_{1,H} \cdot M.$$

All convolution results $T_{j,F}$ with filter H are zero since $T_{i,F}$ for all voxels are equal and $\sum_{i=0}^{k-1} f_{H,i} = 0$ [27].

Stage 3. Column decomposition. Let us calculate the exact values $T_{3,F}$ and errors $E_{3,F}$ of column decomposition with filters L and H.

$$T_{3,LL} = T_{2,L} \cdot S_L, E_{3,LL} = (T_{2,L} + E_{2,L})(S_L + E_{1,L}) - T_{3,LL},$$

$$E_{3,LH} = (T_{2,L} + E_{2,L})E_{1,H}, E_{3,HL} = E_{2,H}(S_L + E_{1,L}), E_{3,HH} = E_{2,H}E_{1,H}.$$

Stage 4. Frame decomposition. Let us calculate the exact values $T_{4,F}$ and errors $E_{4,F}$ of frame decomposition with filters L and H.

$$T_{4,LLL} = T_{3,LL} \cdot S_L, E_{4,LLL} = (T_{3,LL} + E_{3,LL})(S_L + E_{1,L}) - T_{4,LLL},$$

$$E_{4,LLH} = (T_{3,LL} + E_{3,LL})E_{1,H}, E_{4,LHL} = E_{3,LH}(S_L + E_{1,L}),$$

$$E_{4,LHH} = E_{3,LH} \cdot E_{1,H}, E_{4,HLL} = E_{3,HL}(S_L + E_{1,L}), E_{4,HLH} = E_{3,HL} \cdot E_{1,H},$$

$$E_{4,HHL} = E_{3,HH}(S_L + E_{1,L}), E_{4,HHH} = E_{3,HH} \cdot E_{1,H}.$$

Stage 5. Frame reconstruction. Let us calculate the exact values $T_{5,F,l}$ and errors $E_{5,F,l}$ of frame reconstruction with filters L and H, $\varepsilon = 1, 2$.

$$T_{5,LLLL,\varepsilon} = T_{4,LLL} \cdot S_{L,\varepsilon}, E_{5,LLLL,\varepsilon} = (T_{4,LLL} + E_{4,LLL})(S_{L,\varepsilon} + E_{1,L,\varepsilon}) - T_{5,LLLL,\varepsilon},$$

$$E_{5,LLHH,\varepsilon} = E_{4,LLH}(S_{H,\varepsilon} + E_{1,H,\varepsilon}), E_{5,LHLL,\varepsilon} = E_{4,LHL}(S_{L,\varepsilon} + E_{1,L,\varepsilon}),$$

$$E_{5,LHHH,\varepsilon} = E_{4,LHH}(S_{H,\varepsilon} + E_{1,H,\varepsilon}), E_{5,HLLL,\varepsilon} = E_{4,HLL}(S_{L,\varepsilon} + E_{1,L,\varepsilon}),$$

$$E_{5,HLHH,\varepsilon} = E_{4,HLH}(S_{H,\varepsilon} + E_{1,H,\varepsilon}), E_{5,HHLL,\varepsilon} = E_{4,HHL}(S_{L,\varepsilon} + E_{1,L,\varepsilon}),$$

$$E_{5,HHHH,\varepsilon} = E_{4,HHH}(S_{H,\varepsilon} + E_{1,H,\varepsilon}).$$

Stage 6. Frame summation. Let us calculate the errors $E_{6,F,\varepsilon}$ of sums $E_{5,F,\varepsilon}$, $\varepsilon = 1, 2$.

$$E_{6,LL,\varepsilon} = E_{5,LLLL,\varepsilon} + E_{5,LLHH,\varepsilon}, E_{6,LH,\varepsilon} = E_{5,LHLL,\varepsilon} + E_{5,LHHH,\varepsilon},$$

$$E_{6,HL,\varepsilon} = E_{5,HLLL,\varepsilon} + E_{5,HLHH,\varepsilon}, E_{6,HH,\varepsilon} = E_{5,HHLL,\varepsilon} + E_{5,HHHH,\varepsilon}.$$

Stage 7. Column reconstruction. Let us calculate the errors $T_{7,F,\varepsilon}$ and errors $E_{7,F,\varepsilon}$ of column reconstruction with filters L and H.

$$T_{7,LL,1} = T_{5,LLLL,1} \cdot S_{L,1}, T_{7,LL,2} = T_{5,LLLL,2} \cdot S_{L,1}, T_{7,LL,3} = T_{5,LLLL,1} \cdot S_{L,2}, T_{7,LL,4} = T_{5,LLLL,2} \cdot S_{L,2},$$

$$E_{7,LL,1} = (T_{5,LLLL,1} + E_{6,LL,1})(S_{L,1} + E_{1,L,1}) - T_{7,LL,1}, E_{7,LL,2} = (T_{5,LLLL,2} + E_{6,LL,2})(S_{L,1} + E_{1,L,1}) - T_{7,LL,2},$$

$$E_{7,LL,3} = (T_{5,LLLL,1} + E_{6,LL,1})(S_{L,2} + E_{1,L,2}) - T_{7,LL,3}, E_{7,LL,4} = (T_{5,LLLL,2} + E_{6,LL,2})(S_{L,2} + E_{1,L,2}) - T_{7,LL,4},$$

$$E_{7,LH,1} = E_{6,LH,1}(S_{H,1} + E_{1,H,1}), E_{7,LH,2} = E_{6,LH,2}(S_{H,1} + E_{1,H,1}), E_{7,LH,3} = E_{6,LH,1}(S_{H,2} + E_{1,H,2}),$$

$$E_{7,LH,4} = E_{6,LH,2}(S_{H,2} + E_{1,H,2}), E_{7,HL,1} = E_{6,HL,1}(S_{L,1} + E_{1,L,1}), E_{7,HL,2} = E_{6,HL,2}(S_{L,1} + E_{1,L,1}),$$

$$E_{7,HL,3} = E_{6,HL,1}(S_{L,2} + E_{1,L,2}), E_{7,HL,4} = E_{6,HL,2}(S_{L,2} + E_{1,L,2}), E_{7,HH,1} = E_{6,HH,1}(S_{H,1} + E_{1,H,1}),$$

$$E_{7,HH,2} = E_{6,HH,2}(S_{H,1} + E_{1,H,1}), E_{7,HH,3} = E_{6,HH,1}(S_{H,2} + E_{1,H,2}), E_{7,HH,4} = E_{6,HH,2}(S_{H,2} + E_{1,H,2}).$$

Stage 8. Column summation. Let us calculate the errors $E_{8,F,\varepsilon}$ of sums $E_{7,F,\varepsilon}$, $\varepsilon = 1, 2, 3, 4$.

$$E_{8,L,\varepsilon} = E_{7,LL,\varepsilon} + E_{7,LH,\varepsilon}, E_{8,H,\varepsilon} = E_{7,HL,\varepsilon} + E_{7,HH,\varepsilon}.$$

Stage 9. Row reconstruction. Let us calculate the errors $T_{9,\varepsilon}$ and errors $E_{9,F,\varepsilon}$ of column reconstruction with filters L and H.

$$T_{9,1} = T_{7,LL,1} \cdot S_{L,1}, T_{9,2} = T_{7,LL,2} \cdot S_{L,1}, T_{9,3} = T_{7,LL,3} \cdot S_{L,1}, T_{9,4} = T_{7,LL,4} \cdot S_{L,1},$$

$$T_{9,5} = T_{7,LL,1} \cdot S_{L,2}, T_{9,6} = T_{7,LL,2} \cdot S_{L,2}, T_{9,7} = T_{7,LL,3} \cdot S_{L,2}, T_{9,8} = T_{7,LL,4} \cdot S_{L,2},$$

$$E_{9,L,1} = (T_{7,LL,1} + E_{8,L,1})(S_{L,1} + E_{1,L,1}) - T_{9,1}, E_{9,L,2} = (T_{7,LL,2} + E_{8,L,2})(S_{L,1} + E_{1,L,1}) - T_{9,2},$$

$$E_{9,L,3} = (T_{7,LL,3} + E_{8,L,3})(S_{L,1} + E_{1,L,1}) - T_{9,3}, E_{9,L,4} = (T_{7,LL,4} + E_{8,L,4})(S_{L,1} + E_{1,L,1}) - T_{9,4},$$

$$E_{9,L,5} = (T_{7,LL,1} + E_{8,L,1})(S_{L,2} + E_{1,L,2}) - T_{9,5}, E_{9,L,6} = (T_{7,LL,2} + E_{8,L,2})(S_{L,2} + E_{1,L,2}) - T_{9,6},$$

$$E_{9,L,7} = (T_{7,LL,3} + E_{8,L,3})(S_{L,2} + E_{1,L,2}) - T_{9,7}, E_{9,L,8} = (T_{7,LL,4} + E_{8,L,4})(S_{L,2} + E_{1,L,2}) - T_{9,8},$$

$$E_{9,H,1} = E_{8,H,1}(S_{H,1} + E_{1,H,1}), E_{9,H,2} = E_{8,H,2}(S_{H,1} + E_{1,H,1}), E_{9,H,3} = E_{8,H,3}(S_{H,1} + E_{1,H,1}),$$
$$E_{9,H,4} = E_{8,H,4}(S_{H,1} + E_{1,H,1}), E_{9,H,5} = E_{8,H,1}(S_{H,2} + E_{1,H,2}), E_{9,H,6} = E_{8,H,2}(S_{H,2} + E_{1,H,2}),$$
$$E_{9,H,7} = E_{8,H,3}(S_{H,2} + E_{1,H,2}), E_{9,H,8} = E_{8,H,4}(S_{H,2} + E_{1,H,2}).$$

Stage 10. Row summation. Let us calculate the errors $E_{10,\varepsilon}$ of sums $E_{9,F,\varepsilon}$, $\varepsilon = 1,2,3,4,5,6,7,8$.

$$E_{10,\varepsilon} = E_{9,L,\varepsilon} + E_{9,H,\varepsilon}.$$

Stage 11. Normalizing. Let us calculate the errors $E_{11,\varepsilon}$ of rounding downscaled $E_{10,\varepsilon}$ by 2^{-6n}, $\varepsilon = 1,2,3,4,5,6,7,8$.

$$E_{11,\varepsilon} = \lfloor 2^{-6n} E_{10,\varepsilon} \rfloor.$$

The obtained values $E_{11,\varepsilon}$ ($\varepsilon = 1,2,3,4,5,6,7,8$) represent the resulting error of the method and allow for the calculation of the PSNR

$$PSNR = 10\log_{10}\left(8M^2 / \sum_{\varepsilon=1}^{8} E_{11,\varepsilon}^2\right), \tag{5}$$

where $MSE_{grayscale} = MSE_{color} = \frac{1}{8}\sum_{\varepsilon=1}^{8} E_{11,\varepsilon}^2$.

Formula (5) allows determining the minimum quality of a 3D image $db3$, obtained as a result of DWT of the original image I, depending on the maximum brightness and selected bit-width $r = n + 1$ of wavelet filters coefficients $f_{F,i}$.

Calculations results (PSNR, dB) obtained by using our method of wavelet filters coefficients quantizing and final Formula (5) for 3D medical grayscale and color images DWT with various BPC, various bit-width r and numbers $k = 2,4,6,\ldots,20$ of wavelets $db(k/2)$ filters coefficients are presented in Tables 1–3. The cells in bold correspond to the minimum bit-widths of the filter coefficients, at which the processing quality achieves a high level according to the formula (4).

Table 1. Calculation results (PSNR, dB) of 3D medical images (with 8 BPC) DWT by using bit-width r of Daubechies wavelets filters coefficients.

r	db1	db2	db3	db4	db5	db6	db7	db8	db9	db10
10	36.79	36.67	29.87	30.08	31.59	24.60	22.29	24.46	22.19	22.08
11	**44.15**	**43.36**	39.68	34.58	36.67	32.22	28.80	31.17	28.58	27.60
12	57.16	48.71	**44.15**	**41.85**	**43.36**	39.68	34.58	37.82	35.26	35.34
13	∞	∞	51.14	51.14	51.14	**47.16**	**43.18**	**43.36**	**43.36**	39.68
14	∞	∞	∞	∞	∞	57.16	51.14	51.14	51.14	**47.16**
15	∞	∞	∞	∞	∞	∞	∞	∞	∞	∞

Table 2. Calculation results (PSNR, dB) of 3D medical images (with 12 BPC) DWT by using bit-width r of Daubechies wavelets filters coefficients.

r	db1	db2	db3	db4	db5	db6	db7	db8	db9	db10
12	49.43	45.01	42.18	40.05	41.80	38.21	33.76	36.69	34.47	34.40
13	57.46	52.38	49.10	46.76	46.46	44.76	41.11	41.99	41.92	39.07
14	**71.28**	**61.23**	54.39	52.38	52.27	50.35	47.91	47.73	47.65	44.90
15	70.86	71.28	**61.11**	59.15	57.68	57.46	55.52	52.38	53.27	52.27
16	∞	81.28	71.28	**64.37**	**63.79**	**63.79**	**63.15**	57.97	59.85	56.33
17	∞	∞	∞	70.86	75.26	71.28	68.49	**67.30**	**64.37**	**63.79**
18	∞	∞	∞	∞	81.28	81.28	75.26	75.26	75.26	71.28
19	∞	∞	∞	∞	∞	∞	∞	∞	∞	81.28
20	∞	∞	∞	∞	∞	∞	∞	∞	∞	∞

Table 3. Calculation results (*PSNR*, dB) of 3D medical images (with 16 BPC) DWT by using bit-width *r* of Daubechies wavelets filters coefficients.

r	db1	db2	db3	db4	db5	db6	db7	db7	db9	db10
16	77.35	76.55	67.85	63.56	61.94	63.27	61.70	57.52	59.24	55.92
17	**93.90**	77.35	76.91	68.05	71.33	67.85	66.49	65.34	63.50	62.62
18	102.35	**88.73**	80.42	77.35	75.56	73.79	71.69	71.42	71.36	69.35
19	105.36	95.36	85.19	**86.02**	**81.76**	**81.86**	78.48	76.46	78.34	75.56
20	∞	99.34	92.35	91.56	90.05	87.96	**85.19**	**83.84**	**84.87**	**83.46**
21	∞	∞	99.34	105.36	99.34	99.34	92.57	91.38	93.45	87.88
22	∞	∞	∞	∞	105.36	∞	105.36	99.34	105.36	95.36
23	∞	∞	∞	∞	∞	∞	∞	∞	∞	105.36
24	∞	∞	∞	∞	∞	∞	∞	∞	∞	∞

Calculations results (*PSNR*, dB) obtained by using our method of wavelet filters coefficients quantizing and final Formula (5) for 3D medical grayscale and color images DWT with various BPC, various bit-width *r* and numbers $k = 2, 4, 6, \ldots, 20$ of wavelets $sym(k/2)$ filters coefficients are presented in Tables 4–6.

Table 4. Calculation results (*PSNR*, dB) of 3D medical images (with 8 BPC) DWT by using bit-width *r* of symlets filters coefficients.

r	sym1	sym2	sym3	sym4	sym5	sym6	sym7	sym8	sym9	sym10
10	36.79	36.67	29.87	30.08	26.15	26.07	24.53	24.46	21.15	22.08
11	**44.15**	**43.36**	39.68	34.58	32.46	32.22	30.83	31.17	26.75	26.15
12	57.16	48.71	**44.15**	**41.85**	**41.85**	36.99	35.04	37.82	34.71	32.46
13	∞	∞	51.14	51.14	51.14	**43.18**	**44.15**	**43.18**	**43.18**	**41.85**
14	∞	∞	∞	∞	57.16	51.14	57.16	48.71	51.14	47.16
15	∞	∞	∞	∞	∞	∞	∞	∞	∞	∞

Table 5. Calculation results (*PSNR*, dB) of 3D medical images (with 12 BPC) DWT by using bit-width *r* of symlets filters coefficients.

r	sym1	sym2	sym3	sym4	sym5	sym6	sym7	sym8	sym9	sym10
13	57.46	52.38	49.10	46.76	46.46	41.18	42.09	41.03	40.87	39.89
14	**71.28**	**61.23**	54.39	52.38	50.44	49.10	50.46	45.91	47.65	44.90
15	70.86	71.28	**61.11**	**61.23**	57.68	59.46	56.80	51.56	52.38	53.13
16	∞	81.28	71.28	72.82	**63.79**	**67.47**	**63.87**	57.97	58.07	59.37
17	∞	∞	∞	81.28	71.28	75.26	71.28	**64.37**	**65.96**	**67.30**
18	∞	∞	∞	∞	81.28	81.28	81.28	72.82	72.82	75.26
19	∞	∞	∞	∞	∞	∞	∞	∞	∞	∞

Table 6. Calculation results (*PSNR*, dB) of 3D medical images (with 16 BPC) DWT by using bit-width *r* of symlets filters coefficients.

r	sym1	sym2	sym3	sym4	sym5	sym6	sym7	sym8	sym9	sym10
16	77.35	76.55	67.85	70.44	63.48	65.06	63.18	57.52	57.46	58.27
17	**93.90**	77.35	76.91	76.55	67.96	70.97	69.20	63.56	64.39	64.31
18	102.35	**88.73**	80.42	79.43	74.14	73.79	73.90	70.41	70.39	71.33
19	105.36	95.36	85.19	**83.84**	**81.76**	77.89	78.48	77.35	76.46	76.35
20	∞	99.34	92.35	91.56	90.05	**85.19**	**86.02**	**83.84**	**86.02**	**82.16**
21	∞	∞	99.34	105.36	95.36	92.57	92.57	90.05	91.56	90.05
22	∞	∞	∞	∞	105.36	105.36	99.34	96.91	99.34	95.36
23	∞	∞	∞	∞	∞	∞	∞	∞	∞	105.36
24	∞	∞	∞	∞	∞	∞	∞	∞	∞	∞

Calculations results (*PSNR*, dB) obtained by using our method of wavelet filters coefficients quantizing and final Formula (5) for 3D medical grayscale and color images DWT with various BPC, various bit-width r and numbers $k = 6, 12, 18, 24, 30$ of wavelets $coif(k/6)$ filters coefficients are presented in Tables 7–9.

Table 7. Calculation results (*PSNR*, dB) of 3D medical images (with 8 BPC) DWT by using bit-width r of coiflets filters coefficients.

r	*coif*1	*coif*2	*coif*3	*coif*4	*coif*5
11	36.99	29.87	29.48	26.07	25.60
12	**41.85**	36.99	37.82	33.72	32.40
13	48.71	**47.16**	43.18	39.68	37.82
14	∞	51.14	**48.71**	47.16	44.37
15	∞	∞	∞	57.16	57.16
16	∞	∞	∞	∞	∞

Table 8. Calculation results (*PSNR*, dB) of 3D medical images (with 12 BPC) DWT by using bit-width r of coiflets filters coefficients.

r	*coif*1	*coif*2	*coif*3	*coif*4	*coif*5
14	54.39	47.96	45.87	44.09	43.24
15	**61.11**	56.08	52.38	50.35	49.69
16	71.28	**63.79**	58.82	57.46	54.94
17	81.28	75.26	**65.96**	**63.79**	**60.67**
18	∞	81.28	72.82	68.49	68.49
19	∞	∞	∞	75.26	75.26
20	∞	∞	∞	∞	∞

Table 9. Calculation results (*PSNR*, dB) of 3D medical images (with 16 BPC) DWT by using bit-width r of coiflets filters coefficients.

r	*coif*1	*coif*2	*coif*3	*coif*4	*coif*5
17	73.79	70.97	64.39	61.91	59.42
18	**83.54**	73.79	69.56	67.24	66.49
19	87.96	**80.42**	77.35	73.19	72.10
20	92.35	85.19	**84.87**	**80.42**	77.96
21	99.34	95.36	91.56	87.88	**86.02**
22	∞	∞	99.34	95.36	91.38
23	∞	∞	∞	105.36	99.34
24	∞	∞	∞	∞	∞

Let us compile Tables 10–12 based on Tables 1–9 with the minimum values of r, at which the result of 3D medical images DWT with Daubechies wavelets, symlets and coiflets reach a high and maximum quality. For example, the result of 3D medical images (with 8 BPC) DWT with Daubechies wavelet $db8$ reaches high quality at $r = 13$ ($PSNR = 43.36$ dB) and maximum quality at $r = 15$ ($PSNR = \infty$) according to Table 1. The remaining cells are filled in the same way.

Table 10. Minimum values of *r*, at which the result of 3D medical images DWT with Daubechies wavelets reaches high and maximum quality.

BPC	PSNR, dB	db1	db2	db3	db4	db5	db6	db7	db8	db9	db10
8	40	11	11	12	12	12	13	13	13	13	14
	∞	13	13	14	14	14	15	15	15	15	15
12	60	14	14	15	16	16	16	16	17	17	17
	∞	16	17	17	18	19	19	19	19	19	20
16	80	17	18	18	19	19	19	20	20	20	20
	∞	20	21	22	22	23	22	23	23	23	24

Table 11. Minimum values of *r*, at which the result of 3D medical images DWT with symlets reaches high and maximum quality.

BPC	PSNR, dB	sym1	sym2	sym3	sym4	sym5	sym6	sym8	sym9	sym10
8	40	11	11	12	12	12	13	13	13	13
	∞	13	13	14	14	15	15	15	15	15
12	60	14	14	15	15	16	16	17	17	17
	∞	16	17	17	18	19	19	19	19	19
16	80	17	18	18	19	19	20	20	20	20
	∞	20	21	22	22	23	23	23	23	24

Table 12. Minimum values of *r*, at which the result of 3D medical images DWT with coiflets reaches high and maximum quality.

BPC	PSNR, dB	coif1	coif2	coif3	coif4	coif5
8	40	12	13	13	14	14
	∞	14	15	15	16	16
12	60	15	16	17	17	17
	∞	18	19	19	20	20
16	80	18	19	20	20	21
	∞	22	22	23	24	24

We could make the following conclusions based on calculation results presented in the Tables 10–12.

1. Minimum bit-width *r* of wavelet filters coefficients at which the result of 3D medical images with 8 BPC DWT does not contain visible distortions ($PSNR \geq 40$ dB) can be determined by a formula

$$r = 11 + \left\lfloor \sqrt{\frac{k}{2}} \right\rfloor, \qquad (6)$$

where *k* is the number of wavelet filters coefficients.

2. Minimum bit-width *r* of wavelet filters coefficients at which the result of 3D medical images with 12 BPC DWT does not contain visible distortions ($PSNR \geq 60$ dB) can be determined by a formula

$$r = 15 + \left\lfloor \sqrt{\frac{k}{4}} \right\rfloor. \qquad (7)$$

3. Minimum bit-width *r* of wavelet filters coefficients at which the result of 3D medical images with 16 BPC DWT does not contain visible distortions ($PSNR \geq 80$ dB) can be determined by a formula

$$r = 18 + \left\lfloor \sqrt{\frac{k}{3}} \right\rfloor. \qquad (8)$$

4. Minimum bit-width r of wavelet filters coefficients at which the result of 3D medical images DWT does not contain distortions ($PSNR = \infty$) can be determined by a formula

$$r = 5 + B + \left\lfloor \sqrt{\frac{k}{2} - 1} \right\rfloor, \qquad (9)$$

where B is the image BPC.

Formulas (6)–(9) are an approximate since the values r obtained at their use are sometimes redundant, that is, exceed values presented in Tables 10–12. However, they allow one to accurately calculate the non-redundant bit-width of the quantized wavelet filters coefficients in most cases. These formulas are applicable to both grayscale and color images.

3.2. Experiments of the 3D Medical Tomographic Images DWT

The experiments were conducted using MatLab software version R2018b for the three 3D medical tomographic grayscale images: "wmri" is the 8-bit image of size $128 \times 128 \times 27$; "Trufi_COR" is the 12-bit image of size $320 \times 320 \times 30$ and "Body_1.0" is the 16-bit image of size $512 \times 512 \times 507$. These images have the following histograms (Figure 4). The larger the image bitness, the lower its ratio of the average voxel brightness to the maximum allowed. We show the influence of this factor on the image processing quality further.

Images DWT performed as follows: filters coefficients $f_{F,i}$ of the Daubechies wavelets $db(k/2)$ ($k = 2, 4, 6, \ldots, 20$), symlets $sym(k/2)$ ($k = 2, 4, 6, \ldots, 20$) and coiflets $coif(k/6)$ ($k = 6, 12, 18, 24, 30$) were obtained, quantized by multiplying by 2^n ($n = 1, 2, 3, \ldots, 25$) and rounding up according to Formula (2) and converted to fixed-point format; DWT of 3D images implemented; the voxels brightness values of the restored images were scaled by dividing by 2^{6n} and rounding down according to Formula (3) and converted to fixed-point format.

An example of 3D tomographic images "wmri", "Trufi_COR" and "Body_1.0" DWT with wavelet $db8$ is shown in Figures 5–7 respectively. Frames in Figures 6 and 7 are selected to illustrate the error effect on the image processing result. Figures show a gradual improvement in the quality of processing with an increase the bit-width r: in Figures 5b, 6b and 7b visible distortion (Figure 5b is darkened in places, and Figures 6b and 7b are lighted); in Figures 5c, 6c and 7c processed images are indistinguishable by eye from the original images; in Figures 5d, 6d and 7d processed images are identical to the corresponding originals. Experimental results are of higher quality compared with the calculation results. The values $PSNR = 47.11$ dB and $PSNR = \infty$ at $r = 12$ and $r = 15$ respectively (Figure 5) obtained after 8-bit image "wmri" DWT with wavelet $db8$ exceed the corresponding calculated values $PSNR = 37.82$ dB and $PSNR = \infty$ at $r = 12$ and $r = 15$ respectively (Table 1). The values $PSNR = 64.57$ dB and $PSNR = \infty$ at $r = 12$ and $r = 17$ respectively (Figure 6) obtained after 12-bit image "Trufi_COR" DWT with wavelet $db8$ exceed the corresponding calculated values $PSNR = 36.67$ dB, $PSNR = 67.30$ dB at $r = 12$ and $r = 17$ respectively (Table 2). Similarly, for "Body_1.0".

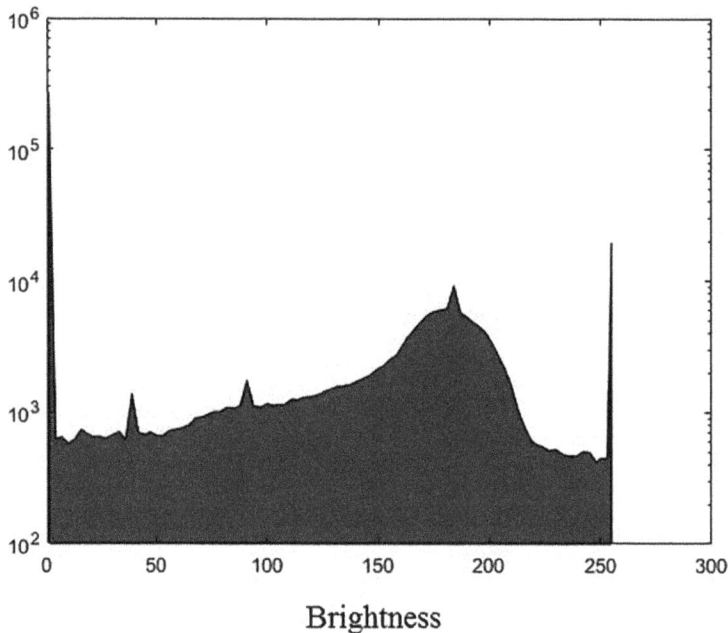

(a)

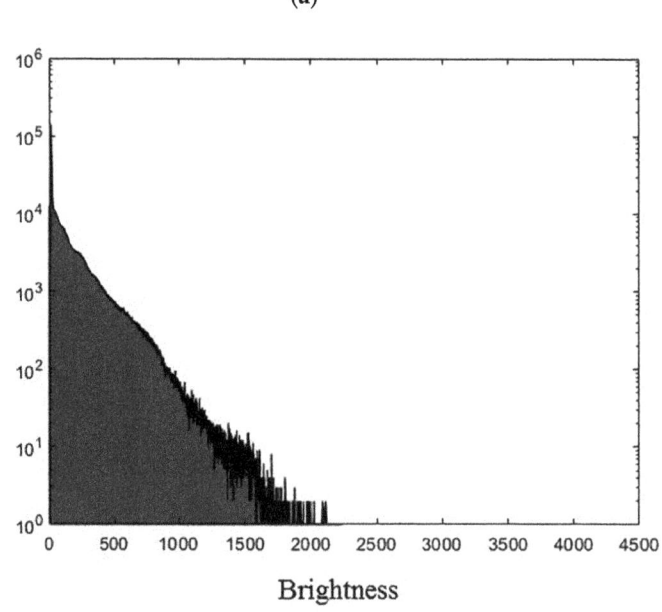

(b)

Figure 4. *Cont.*

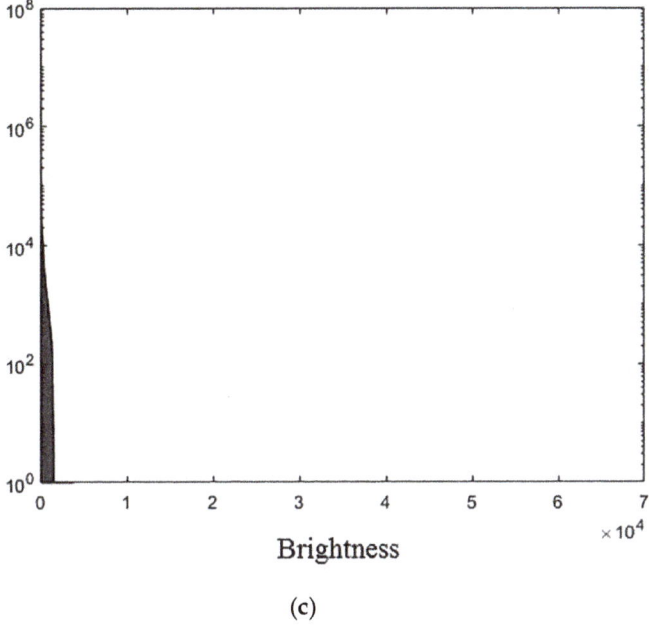

(c)

Figure 4. Histograms of used images: (**a**) "wmri", average brightness 63.276; (**b**) "Trufi_COR", average brightness 129.796 and (**c**) "Body_1.0", average brightness 21.053.

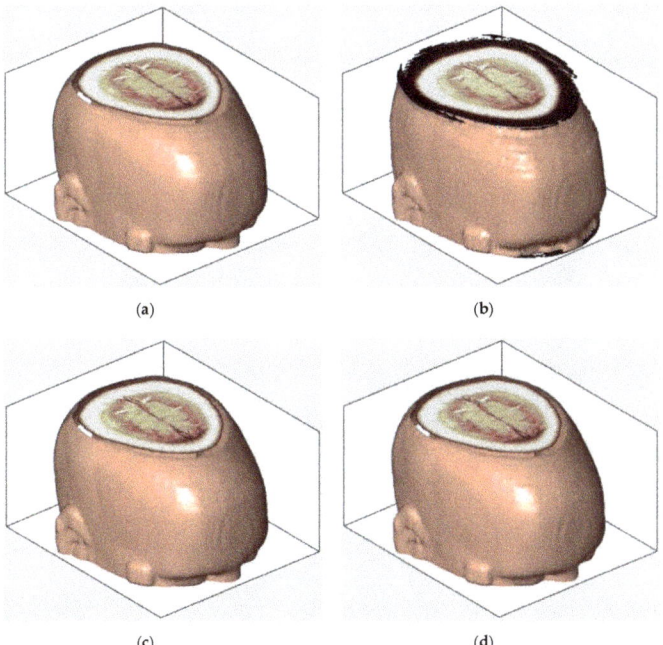

Figure 5. Example of 3D tomographic 8-bit image "wmri" DWT by *db*8 wavelet: (**a**) original image; processed image: (**b**) $r = 9$, $PSNR = 27.62$ dB; (**c**) $r = 12$, $PSNR = 47.11$ dB and (**d**) $r = 15$, $PSNR = \infty$.

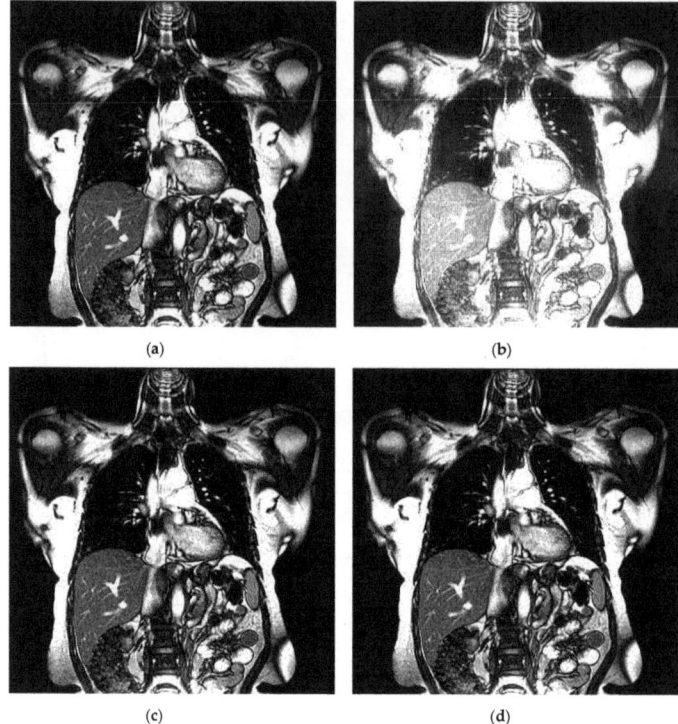

Figure 6. Example of 3D tomographic 12-bit image "Trufi_COR" (15-th frame) DWT by $db8$ wavelet: (**a**) original image; processed image: (**b**) $r = 7$, $PSNR = 30.27$ dB; (**c**) $r = 12$, $PSNR = 64.57$ dB and (**d**) $r = 17$, $PSNR = \infty$.

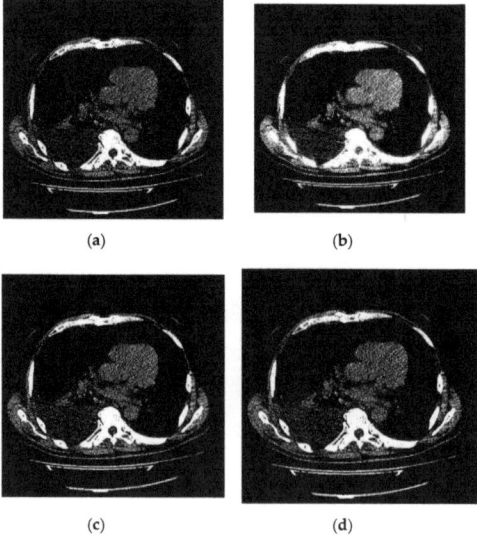

Figure 7. Example of 3D tomographic 16-bit image "Body_1.0" (1-st frame) DWT by $db8$ wavelet: (**a**) original image; processed image: (**b**) $r = 7$, $PSNR = 64.05$ dB; (**c**) $r = 10$, $PSNR = 85.30$ dB and (**d**) $r = 17$, $PSNR = \infty$.

The image processing results were analyzed using *PSNR* and structure similarity (*SSIM*) [52], calculating by formula

$$SSIM(I,\widetilde{I}) = \frac{(2\mu_I\mu_{\widetilde{I}} + c_1)(2\sigma_{I\widetilde{I}} + c_2)}{(\mu_I^2 + \mu_{\widetilde{I}}^2 + c_1)(\sigma_I^2 + \sigma_{\widetilde{I}}^2 + c_2)},$$

where: μ_I is the average of I; $\mu_{\widetilde{I}}$ is the average of $\widetilde{I}$; σ_I^2 is the variance of I; $\sigma_{\widetilde{I}}^2$ is the variance of $\widetilde{I}$; $c_1 = (0.01 \cdot M)^2$; $c_2 = (0.03 \cdot M)^2$ and M is the maximum brightness of the image voxels. Experimental results (*PSNR*, dB; *SSIM*) of DWT of 3D tomographic grayscale images "wmri" (8-bit), "Trufi_COR" (12-bit) and "Body_1.0" (16-bit) for various bit-width r and numbers $k = 2, 4, 6, \ldots, 20$ of wavelets $db(k/2)$ filters coefficients are presented in Tables 13–18. The cells in bold correspond to the minimum bit-widths of the filter coefficients, at which the processing quality achieves a high level according to the formula (4).

Table 13. Experimental results (*PSNR*, dB) of 3D tomographic 8-bit image "wmri" DWT by using bit-width r of Daubechies wavelets filters coefficients.

r	db1	db2	db3	db4	db5	db6	db7	db8	db9	db10
9	37.77	37.51	31.59	31.45	33.45	27.81	25.33	27.62	24.16	25.03
10	**44.78**	**44.47**	38.16	37.95	40.21	32.98	30.81	32.95	30.76	30.67
11	52.66	52.44	**48.29**	42.84	**44.97**	41.03	37.43	39.59	37.37	36.48
12	69.55	56.29	53.32	50.66	52.88	48.64	**43.25**	**47.11**	**44.52**	**44.44**
13	∞	∞	70.86	60.74	60.36	56.92	52.47	53.42	53.32	50.09
14	∞	∞	∞	∞	∞	93.45	65.73	64.80	64.10	57.65
15	∞	∞	∞	∞	∞	∞	∞	∞	∞	∞

Table 14. Experimental results (*SSIM*) of 3D tomographic 8-bit image "wmri" DWT by using bit-width r of Daubechies wavelets filters coefficients.

r	db1	db2	db3	db4	db5	db6	db7	db8	db9	db10
9	0.9998	0.9998	0.9991	0.9990	0.9993	0.9975	0.9953	0.9969	0.9936	0.9943
10	1.0000	1.0000	0.9998	0.9998	0.9998	0.9993	0.9987	0.9991	0.9985	0.9984
11	1.0000	1.0000	1.0000	0.9999	0.9999	0.9999	0.9997	0.9998	0.9997	0.9996
12	1.0000	1.0000	1.0000	1.0000	1.0000	1.0000	0.9999	1.0000	0.9999	0.9999
13	1.0000	1.0000	1.0000	1.0000	1.0000	1.0000	1.0000	1.0000	1.0000	1.0000

Table 15. Experimental results (*PSNR*, dB) of 3D tomographic 12-bit image "Trufi_COR" DWT by using bit-width r of Daubechies wavelets filters coefficients.

r	db1	db2	db3	db4	db5	db6	db7	db8	db9	db10
9	55.35	55.15	49.64	49.52	51.56	46.10	43.74	46.08	42.62	43.54
10	**61.94**	**61.74**	55.93	55.74	58.14	51.13	49.08	51.29	49.14	49.10
11	69.03	69.02	**65.33**	**60.46**	**62.66**	58.84	55.49	57.68	55.53	54.82
12	77.25	72.30	69.89	67.43	69.64	**65.88**	**61.00**	**64.57**	**62.24**	**62.33**
13	88.60	81.40	78.04	75.10	75.33	73.56	69.22	70.48	70.39	67.51
14	118.20	96.42	85.23	82.51	82.65	80.99	77.38	77.57	77.62	74.43
15	129.34	119.88	99.48	93.90	90.03	91.05	88.62	84.20	85.12	84.30
16	∞	∞	∞	110.99	105.34	109.92	105.08	94.67	98.96	91.61
17	∞	∞	∞	∞	∞	∞	∞	∞	121.21	117.93
18	∞	∞	∞	∞	∞	∞	∞	∞	∞	∞

Table 16. Experimental results (*SSIM*) of 3D tomographic 12-bit image "Trufi_COR" DWT by using bit-width *r* of Daubechies wavelets filters coefficients.

r	db1	db2	db3	db4	db5	db6	db7	db8	db9	db10
9	0.9996	0.9995	0.9981	0.9979	0.9986	0.9948	0.9902	0.9939	0.9864	0.9884
10	0.9999	0.9999	0.9996	0.9996	0.9997	0.9985	0.9972	0.9982	0.9968	0.9967
11	1.0000	1.0000	1.0000	0.9999	0.9999	0.9998	0.9994	0.9996	0.9993	0.9991
12	1.0000	1.0000	1.0000	1.0000	1.0000	1.0000	0.9999	0.9999	0.9999	0.9999
13	1.0000	1.0000	1.0000	1.0000	1.0000	1.0000	1.0000	1.0000	1.0000	1.0000

Table 17. Experimental results (*PSNR*, dB) of 3D tomographic 16-bit image "Body_1.0" DWT by using bit-width *r* of Daubechies wavelets filters coefficients.

r	db1	db2	db3	db4	db5	db6	db7	db8	db9	db10
7	77.99	77.69	70.86	68.90	68.96	65.38	65.39	64.05	62.04	59.98
8	**87.98**	81.65	78.77	76.46	76.54	71.53	70.89	72.11	68.08	68.88
9	88.22	88.00	**82.85**	**82.90**	**84.95**	79.80	77.67	79.97	76.69	77.63
10	94.84	94.62	89.15	88.98	91.67	**84.80**	83.03	**85.30**	**83.27**	**83.27**
11	101.92	101.93	98.64	93.87	96.45	92.39	89.47	91.69	89.76	89.16
12	109.95	105.14	103.17	100.98	103.38	99.76	94.87	98.86	96.63	96.80
13	120.74	114.03	111.17	108.38	108.91	107.49	103.36	104.87	104.83	102.16
14	167.56	127.71	118.07	115.86	116.02	115.25	111.44	111.96	112.20	109.24
15	166.77	169.11	131.41	126.83	122.99	125.99	123.36	119.26	119.89	119.63
16	∞	∞	∞	144.12	138.29	145.66	140.29	130.67	135.37	128.20
17	∞	∞	∞	∞	∞	∞	∞	∞	172.79	161.13
18	∞	∞	∞	∞	∞	∞	∞	∞	∞	∞

Table 18. Experimental results (*SSIM*) of 3D tomographic 16-bit image "Body_1.0" DWT by using bit-width *r* of Daubechies wavelets filters coefficients.

r	db1	db2	db3	db4	db5	db6	db7	db8	db9	db10
7	0.9999	0.9999	0.9996	0.9994	0.9994	0.9987	0.9986	0.9982	0.9970	0.9953
8	1.0000	1.0000	0.9999	0.9999	0.9999	0.9997	0.9996	0.9997	0.9992	0.9993
9	1.0000	1.0000	1.0000	1.0000	1.0000	0.9999	0.9999	0.9999	0.9999	0.9999
10	1.0000	1.0000	1.0000	1.0000	1.0000	1.0000	1.0000	1.0000	1.0000	1.0000

Experimental results (*PSNR*, dB; *SSIM*) of DWT of 3D tomographic grayscale images "wmri" (8-bit), "Trufi_COR" (12-bit) and "Body_1.0" (16-bit) for various bit-width *r* and numbers $k = 2, 4, 6, \ldots, 20$ of wavelets $sym(k/2)$ filters coefficients are presented in Tables 19–24.

Table 19. Experimental results (*PSNR*, dB) of 3D tomographic 8-bit image "wmri" DWT by using bit-width *r* of symlets filters coefficients.

r	sym1	sym2	sym3	sym4	sym5	sym6	sym7	sym8	sym9	sym10
9	37.77	37.51	31.59	31.35	27.75	29.25	26.23	25.01	22.24	24.91
10	**44.78**	**44.47**	38.16	38.01	34.33	34.16	32.84	32.72	29.60	30.54
11	52.66	52.44	**48.29**	42.69	**41.11**	**41.00**	39.51	39.38	35.37	34.59
12	69.55	56.29	53.32	50.35	50.34	45.38	**44.43**	46.72	43.12	41.43
13	∞	∞	70.86	60.35	60.06	52.29	53.30	52.18	52.18	50.82
14	∞	∞	∞	∞	87.43	74.66	79.40	59.06	63.68	57.27
15	∞	∞	∞	∞	∞	∞	∞	∞	∞	∞

Table 20. Experimental results (*SSIM*) of 3D tomographic 8-bit image "wmri" DWT by using bit-width *r* of symlets filters coefficients.

r	sym1	sym2	sym3	sym4	sym5	sym6	sym7	sym8	sym9	sym10
9	0.9998	0.9998	0.9991	0.9990	0.9979	0.9982	0.9967	0.9954	0.9917	0.9944
10	1.0000	1.0000	0.9998	0.9998	0.9995	0.9995	0.9992	0.9992	0.9984	0.9985
11	1.0000	1.0000	1.0000	0.9999	0.9999	0.9999	0.9998	0.9998	0.9996	0.9995
12	1.0000	1.0000	1.0000	1.0000	1.0000	1.0000	0.9999	1.0000	0.9999	0.9999
13	1.0000	1.0000	1.0000	1.0000	1.0000	1.0000	1.0000	1.0000	1.0000	1.0000

Table 21. Experimental results (*PSNR*, dB) of 3D tomographic 12-bit image "Trufi_COR" DWT by using bit-width *r* of symlets filters coefficients.

r	sym1	sym2	sym3	sym4	sym5	sym6	sym7	sym8	sym9	sym10
9	55.35	55.15	49.64	49.35	45.83	47.40	44.46	43.29	40.55	43.33
10	61.94	61.74	55.93	55.82	52.32	52.11	50.96	50.87	47.78	48.86
11	69.03	69.02	65.33	60.22	58.78	58.74	57.42	57.33	53.42	52.74
12	77.25	72.30	69.89	67.03	67.23	62.83	61.88	64.07	60.87	59.29
13	88.60	81.40	78.04	74.82	75.01	68.86	70.12	68.84	68.94	67.90
14	118.20	96.42	85.23	82.14	80.24	78.63	80.16	74.73	76.99	73.83
15	129.34	119.88	99.48	97.83	90.32	93.04	89.60	82.15	83.43	84.56
16	∞	∞	∞	∞	109.58	115.30	107.15	93.49	93.67	94.62
17	∞	∞	∞	∞	∞	∞	∞	117.03	119.88	119.49
18	∞	∞	∞	∞	∞	∞	∞	∞	∞	∞

Table 22. Experimental results (*SSIM*) of 3D tomographic 12-bit image "Trufi_COR" DWT by using bit-width *r* of symlets filters coefficients.

r	sym1	sym2	sym3	sym4	sym5	sym6	sym7	sym8	sym9	sym10
9	0.9996	0.9995	0.9981	0.9980	0.9955	0.9964	0.9932	0.9905	0.9825	0.9886
10	0.9999	0.9999	0.9996	0.9996	0.9990	0.9989	0.9984	0.9983	0.9966	0.9968
11	1.0000	1.0000	1.0000	0.9999	0.9998	0.9998	0.9997	0.9997	0.9991	0.9988
12	1.0000	1.0000	1.0000	1.0000	1.0000	0.9999	0.9999	0.9999	0.9998	0.9998
13	1.0000	1.0000	1.0000	1.0000	1.0000	1.0000	1.0000	1.0000	1.0000	1.0000

Table 23. Experimental results (*PSNR*, dB) of 3D tomographic 16-bit image "Body_1.0" DWT by using bit-width *r* of symlets filters coefficients.

r	sym1	sym2	sym3	sym4	sym5	sym6	sym7	sym8	sym9	sym10
7	77.99	77.69	70.86	66.59	68.20	66.55	62.15	62.64	61.78	61.58
8	87.98	81.65	78.77	74.26	74.19	74.41	71.33	70.73	68.46	69.41
9	88.22	88.00	82.85	82.52	79.09	80.85	77.89	76.90	74.25	77.28
10	94.84	94.62	89.15	89.03	85.67	85.47	84.57	84.48	81.44	82.83
11	101.92	101.93	98.64	93.36	92.14	92.21	91.01	90.95	87.22	86.66
12	109.95	105.14	103.17	100.15	100.63	96.34	95.53	97.87	94.76	93.33
13	120.74	114.03	111.17	107.82	108.37	102.48	103.73	102.57	102.98	102.12
14	167.56	127.71	118.07	114.95	113.49	112.18	113.78	108.50	111.20	108.03
15	166.77	169.11	131.41	129.63	123.57	126.81	123.10	115.96	117.59	118.94
16	∞	∞	∞	∞	142.56	147.12	140.64	127.84	128.64	129.62
17	∞	∞	∞	∞	∞	∞	∞	151.00	160.08	159.78
18	∞	∞	∞	∞	∞	∞	∞	∞	∞	∞

Table 24. Experimental results (*SSIM*) of 3D tomographic 16-bit image "Body_1.0" DWT by using bit-width *r* of symlets filters coefficients.

r	sym1	sym2	sym3	sym4	sym5	sym6	sym7	sym8	sym9	sym10
7	0.9999	0.9999	0.9996	0.9991	0.9994	0.9990	0.9974	0.9976	0.9970	0.9969
8	1.0000	1.0000	0.9999	0.9998	0.9998	0.9998	0.9997	0.9996	0.9993	0.9994
9	1.0000	1.0000	1.0000	1.0000	0.9999	1.0000	0.9999	0.9999	0.9998	0.9999
10	1.0000	1.0000	1.0000	1.0000	1.0000	1.0000	1.0000	1.0000	1.0000	1.0000

Experimental results (*PSNR*, dB; *SSIM*) of DWT of 3D tomographic grayscale images "wmri" (8-bit), "Trufi_COR" (12-bit) and "Body_1.0" (16-bit) for various bit-width *r* and numbers $k = 6, 12, 18, 24, 30$ of wavelets $coif(k/6)$ filters coefficients are presented in Tables 25–30.

Table 25. Experimental results (*PSNR*, dB) of 3D tomographic 8-bit image "wmri" DWT by using bit-width *r* of coiflets filters coefficients.

r	coif1	coif2	coif3	coif4	coif5
9	33.91	27.72	24.02	20.62	21.25
10	**40.58**	34.26	30.54	27.99	27.35
11	44.93	38.35	38.25	34.55	35.28
12	50.59	**45.41**	**46.59**	**43.00**	**41.96**
13	61.35	55.11	52.12	48.07	48.21
14	∞	66.81	58.76	56.43	56.52
15	∞	∞	∞	99.82	96.81
16	∞	∞	∞	∞	∞

Table 26. Experimental results (*SSIM*) of 3D tomographic 8-bit image "wmri" DWT by using bit-width *r* of coiflets filters coefficients.

r	coif1	coif2	coif3	coif4	coif5
9	0.9995	0.9976	0.9940	0.9870	0.9873
10	0.9999	0.9995	0.9987	0.9973	0.9966
11	1.0000	0.9998	0.9998	0.9994	0.9994
12	1.0000	1.0000	1.0000	0.9999	0.9999
13	1.0000	1.0000	1.0000	1.0000	1.0000

Table 27. Experimental results (*PSNR*, dB) of 3D tomographic 12-bit image "Trufi_COR" DWT by using bit-width *r* of coiflets filters coefficients.

r	coif1	coif2	coif3	coif4	coif5
10	58.14	52.30	48.76	46.38	45.24
11	**62.21**	56.20	56.34	52.78	53.00
12	67.16	**62.90**	**64.06**	**60.82**	59.41
13	75.02	71.54	68.84	65.25	**64.93**
14	85.17	77.12	74.73	72.81	72.40
15	99.11	87.99	83.32	81.23	80.64
16	126.33	103.94	95.26	92.00	88.75
17	∞	∞	121.68	109.44	101.44
18	∞	∞	∞	∞	∞

Table 28. Experimental results (*SSIM*) of 3D tomographic 12-bit image "Trufi_COR" DWT by using bit-width *r* of coiflets filters coefficients.

r	coif1	coif2	coif3	coif4	coif5
10	0.9998	0.9989	0.9972	0.9944	0.9918
11	0.9999	0.9996	0.9995	0.9988	0.9986
12	1.0000	0.9999	0.9999	0.9998	0.9997
13	1.0000	1.0000	1.0000	0.9999	0.9999
14	1.0000	1.0000	1.0000	1.0000	1.0000

Table 29. Experimental results (*PSNR*, dB) of 3D tomographic 16-bit image "Body_1.0" DWT by using bit-width r of coiflets filters coefficients.

r	coif1	coif2	coif3	coif4	coif5
8	78.19	71.20	70.48	65.46	64.87
9	**84.76**	79.33	76.10	73.01	73.31
10	91.16	**85.79**	**82.46**	80.39	79.40
11	95.23	89.68	90.16	86.87	**87.29**
12	100.21	96.41	97.92	94.99	93.71
13	107.91	105.11	102.80	99.41	99.41
14	117.81	110.61	108.67	107.14	107.07
15	130.24	121.55	117.33	115.73	115.45
16	171.54	136.59	129.87	128.02	124.92
17	∞	∞	168.53	146.89	138.98
18	∞	∞	∞	∞	∞

Table 30. Experimental results (*SSIM*) of 3D tomographic 16-bit image "Body_1.0" DWT by using bit-width r of coiflets filters coefficients.

r	coif1	coif2	coif3	coif4	coif5
8	0.9999	0.9997	0.9996	0.9986	0.9983
9	1.0000	0.9999	0.9999	0.9997	0.9997
10	1.0000	1.0000	1.0000	1.0000	0.9999
11	1.0000	1.0000	1.0000	1.0000	1.0000

Calculation results from Tables 10–12 supplemented by experimental results from Tables 13–30 and the difference between them is presented in Tables 31–33.

Table 31. Minimum values of r, at which the result of 3D tomographic images DWT by Daubechies wavelets reaches high and maximum quality.

BPC	PSNR, dB	Results	db1	db2	db3	db4	db5	db6	db8	db10
8	40	Calculation	11	11	12	12	12	13	13	14
		Experimental	10	10	11	11	10	11	12	12
		Difference	1	1	1	1	2	2	1	2
	∞	Calculation	13	13	14	14	14	15	15	15
		Experimental	13	13	14	14	14	15	15	15
		Difference	0	0	0	0	0	0	0	0
12	60	Calculation	14	14	15	16	16	16	17	17
		Experimental	10	10	11	11	11	12	12	12
		Difference	4	4	4	5	5	4	5	5
	∞	Calculation	16	17	17	18	19	19	19	20
		Experimental	16	16	16	17	17	17	17	18
		Difference	0	1	1	1	2	2	2	2
16	80	Calculation	17	18	18	19	19	19	20	20
		Experimental	8	8	9	9	9	10	10	10
		Difference	9	10	9	10	10	9	10	10
	∞	Calculation	20	21	22	22	23	22	23	24
		Experimental	16	16	16	17	17	17	17	18
		Difference	4	5	6	5	6	5	6	6

Table 32. Minimum values of *r*, at which the result of 3D tomographic images DWT by symlets reaches high and maximum quality.

BPC	PSNR, dB	Results	sym1	sym2	sym4	sym6	sym8	sym10
8	40	Calculation	11	11	12	13	13	13
		Experimental	10	10	11	11	12	12
		Difference	1	1	1	2	1	1
	∞	Calculation	13	13	14	15	15	15
		Experimental	13	13	14	15	15	15
		Difference	0	0	0	0	0	0
12	60	Calculation	14	14	15	16	17	17
		Experimental	10	10	11	12	12	13
		Difference	4	4	4	4	5	4
	∞	Calculation	16	17	18	19	19	19
		Experimental	16	16	16	17	18	18
		Difference	0	1	2	2	1	1
16	80	Calculation	17	18	19	20	20	20
		Experimental	8	8	9	9	10	10
		Difference	9	10	10	11	10	10
	∞	Calculation	20	21	22	23	23	24
		Experimental	16	16	16	17	18	18
		Difference	4	5	6	6	5	6

Table 33. Minimum values of *r*, at which the result of 3D tomographic images DWT by coiflets reaches high and maximum quality.

BPC	PSNR, dB	Results	coif1	coif2	coif3	coif4	coif5
8	40	Calculation	12	13	13	14	14
		Experimental	10	12	12	12	12
		Difference	2	1	1	2	2
	∞	Calculation	14	15	15	16	16
		Experimental	14	15	15	16	16
		Difference	0	0	0	0	0
12	60	Calculation	15	16	17	17	17
		Experimental	11	12	12	12	13
		Difference	4	4	5	5	4
	∞	Calculation	18	19	19	20	20
		Experimental	17	17	18	18	18
		Difference	1	2	1	2	2
16	80	Calculation	18	19	20	20	21
		Experimental	9	10	10	10	11
		Difference	9	9	10	10	10
	∞	Calculation	22	22	23	24	24
		Experimental	17	17	18	18	18
		Difference	5	5	5	6	6

Experimental results (*PSNR*, dB) of various 3D tomographic 12-bit grayscale images DWT by wavelet *db*4 with bit-width $r = 11$ of filters coefficients are presented in Table 34 and in Figure 8.

Table 34. Experimental results (*PSNR*, dB) of 3D tomographic images DWT by wavelet *db*4 with bit-width $r = 11$ of filters coefficients.

Image Name	Average Brightness	PSNR, dB
SUB_1st pass	16.89	74.57
cor shared echo_SUB_MIP_COR	33.92	72.87
MIP thin cor first phase	55.16	67.63
mra highres.ce_S47_DIS2D	63.74	69.07
cor thin mips ist pass	67.92	64.58
mra highres.ce_S48_DIS2D	77.29	67.32
sag timing run-flash_MIP_SAG	91.71	62.81
cine_retro_normal_lvot	109.46	63.07
cine_retro_normal_rvot	123.87	60.63
Trufi_COR	129.80	60.46
Trufi_SAG	130.79	59.97
cine_retro_normal_sa	133.50	60.17
cine_retro_normal_lvla	134.35	60.41
cine_retro_normal_hla	144.48	59.72
cine_retro_aortic valve	157.94	58.87
Trufi_TRANS	162.25	58.83
t1_fl2d_cor_pre-post	187.42	58.39

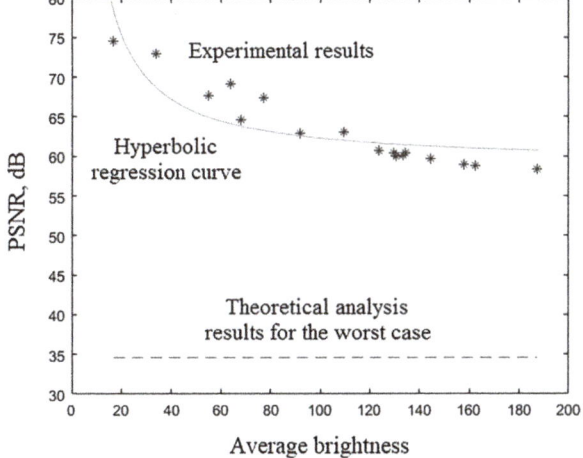

Figure 8. Experimental results of 3D tomographic 12-bit images DWT by wavelet *db*4 with bit-width $r = 11$ of filters coefficients.

The nonlinear hyperbolic regression [53] curve for the data from Table 34 was plotted in Figure 8 and has the equation $PSNR = 58.98 + 328.78/A$, where A is the average brightness of the image voxels. The F-test value [54] for constructed nonlinear hyperbolic regression curve is $F = 42.24$ actually observed. The F-test critical value [55] for false-rejection probability 0.001 with degrees of freedom $k_1 = p - 1 = 2 - 1 = 1$ and $k_2 = m - p = 17 - 2 = 15$ is $F_{0.001;1,15} = 16.59$, where p is the regression equation estimated parameters number and m is the images number from Table 34. Since $F > F_{0.001;1,15}$ resulting regression equation is significant at false-rejection probability 0.001. Equation asymptote exceeds the corresponding theoretical calculations values.

4. Discussion

Experimental results, the main of which are presented in Tables 31–33, show that all *PSNR* values obtained as calculation results were not bigger than the *PSNR* values obtained as experimental results. This confirms the accuracy of theoretical analysis. Thus, the derived Formulas (6)–(9) could be used for determining the minimum bit-width of wavelet filters coefficients, at which the result of 3D medical tomographic images DWT reaches high ($PSNR \geq 40$ dB for images with 8 BPC, $PSNR \geq 60$ dB for images with 12 BPC and $PSNR \geq 80$ dB for images with 16 BPC according to Formula (4)) and maximum ($PSNR = \infty$) quality respectively. Tables 13–30 show that $SSIM$ values obtained as a calculation result were set to one when using 4 decimal places in simulating 8-, 12- and 16-bit images when the $PSNR$ was approximately 45, 65 and 80 dB, respectively. Thus, both $PSNR$ and $SSIM$ metrics used confirm high-quality image processing. The experiment of 3D 8-bit medical tomographic image DWT required 1–2 bits less for wavelet filters coefficients than the calculations require for high-quality processing since the worst case was predicted in theoretical analysis. An even greater decrease in the bit-width of wavelet filter coefficients led to even greater savings in hardware resources. The difference between the obtained theoretical and experimental values increased significantly in the case of 12-bit and 16-bit images. The 12-bit tomographic image required 4–5 bits and 1–2 bits less for wavelet filters coefficients to achieve high and maximum processing quality respectively. This difference increased to 9–10 and 5–6 bits respectively in the case of a 16-bit image. This is because the range of voxel brightness values significantly increased in 12- and 16-bit images. The average brightness of the image voxels varied insignificantly at this time (was within the 8-bit range) since the high-order bits were rarely used. Thus, the ratio of the average voxel brightness to the maximum allowable value of M decreased with increasing BPC of images, which were demonstrated by histograms in Figure 4. This led to much faster achievement of high and maximum quality compared with the theoretical analysis results.

The darkening and lighting in Figures 5–7 were due to the low accuracy of wavelet filters coefficients quantization used for image processing. The excessive character of quantization error led to an increase in the voxels brightness values of the processed images. Figures 6b and 7b turned out to be lighted since 12- and 16-bit images had a brightness margin, which is shown by the histograms in Figure 4b,c. However, the range of brightness values of the 8-bit image was fully utilized (Figure 4a) and the quantization error led to the computational range overflow. The voxels brightness values that exceeded the range went to zero as a result of this.

Table 34 and Figure 8 show the dependence of the 12-bit medical tomographic images processing quality of their average voxels brightness. This dependence had a nonlinear hyperbolic regression form. Equation asymptote exceeded the corresponding theoretical calculation values. The processing quality by $PSNR$ metric (from 74.57 to 58.39) decreased with an increase in the average voxels brightness (from 16.89 to 187.42). The difference in the image processing quality with the minimum and maximum values of the average brightness according to Table 34 was more than 15 dB. It was commensurate with the difference in the processing quality of the same image by the same wavelet with filter coefficients bit-width that differ by two, according to Tables 15, 21 and 27. That is, we would need 2 bits less for wavelet filter coefficients for high-quality processing of a 12-bit image with an average brightness of 16.89 than for processing a 12-bit image with an average brightness of 187.42. The average voxels brightness of the medical image can vary in different ranges depending on many factors: from the medical image modalities; from the analyzing device type; from specific device settings; from the analyzed organ or group of organs; etc. Thus, the requirements for the digit capacity of wavelet filter coefficients can be relaxed, depending on the ability to take into account many factors related to the nature of the images obtained as a result of medical tests. Summarizing, the quality of 3D medical tomographic images DWT primarily depends on their bits per color, on average voxels brightness, on the number of wavelet filters coefficients and to a lesser extent on the type of wavelet.

Minimum bit-width r of wavelet filters coefficients for 3D medical tomographic images DWT is defined as follows: determine BPC of images (for example, 8, 12 or 16 BPC); select a quality threshold of image processing (for example, $PSNR = 40$ dB, $PSNR = 60$ dB, $PSNR = 80$ dB or $PSNR = \infty$); choose the wavelet with the number of coefficients k; calculate bit-width r of wavelet filters coefficients by Formulas (5)–(9) depending on the quality threshold of image processing selected.

5. Conclusions

The problem of analyzing the quantization noise effect in coefficients of DWT filters for 3D medical imaging was solved. The method was proposed for wavelet filters coefficients quantizing, which allows minimizing resources in hardware implementation. The method was developed for estimating the maximum error of 3D grayscale and color images DWT with various BPC. The derived Formula (5) allows determining the minimum quality of 3D medical images DWT depending on the wavelet used, bit-width of wavelet filters coefficients and BPC. We proved that Formulas (6)–(9) can be used to determine the minimum bit-width of wavelet filters coefficients for which the result of 3D images DWT reaches high ($PSNR \geq 40$ dB for images with 8 BPC, $PSNR \geq 60$ dB for images with 12 BPC and $PSNR \geq 80$ dB for images with 16 BPC) and maximum ($PSNR = \infty$) quality respectively depending on the wavelet used. The experiments of the 3D tomographic images DWT showed that the bit-width of wavelet filters coefficients could be significantly reduced for high-quality medical imaging compared to theoretical analysis results. All data were presented in a fixed-point format and rounding operations were simplified in the proposed method of 3D images DWT.

The proposed DWT method could be used in a wide range of applications for denoising and compression of 3D medical images. Given the need to improve the efficiency of medical visual data processing methods, further research can be expected in this direction.

Author Contributions: Conceptualization, P.L.; Data curation, P.L.; Formal analysis, N.C.; Investigation, N.N. and P.L.; Methodology, N.C.; Project administration, N.C.; Resources, N.C.; Software, N.N.; Supervision, N.C.; Validation, N.C.; Visualization, N.N.; Writing—original draft, N.N. and P.L.; Writing, review & editing, N.N., P.L. and N.C. All authors have read and agreed to the published version of the manuscript.

Funding: This research was funded by the Russian Foundation for Basic Research (RFBR), grants numbers 19-07-00130 A and 18-37-20059 mol-a-ved, and the Council on grants of the President of the Russian Federation, grant number SP-2245.2018.5.

Acknowledgments: We are thankful to the Stavropol Regional Clinical Advisory and Diagnostic Center for providing tomographic images.

Conflicts of Interest: The authors declare no conflict of interest.

References

1. Barnhill, E.; Hollis, L.; Sack, I.; Braun, J.; Hoskins, P.R.; Pankaj, P.; Brown, C.; van Beek, E.J.R.; Roberts, N. Nonlinear multiscale regularisation in MR elastography: Towards fine feature mapping. *Med. Image Anal.* **2017**, *35*, 133–145. [CrossRef]
2. Benou, A.; Veksler, R.; Friedman, A.; Riklin Raviv, T. Ensemble of expert deep neural networks for spatio-temporal denoising of contrast-enhanced MRI sequences. *Med. Image Anal.* **2017**, *42*, 145–159. [CrossRef]
3. Lai, Z.; Qu, X.; Liu, Y.; Guo, D.; Ye, J.; Zhan, Z.; Chen, Z. Image reconstruction of compressed sensing MRI using graph-based redundant wavelet transform. *Med. Image Anal.* **2016**, *27*, 93–104. [CrossRef]
4. Lucas, L.F.R.; Rodrigues, N.M.M.; Da Silva Cruz, L.A.; De Faria, S.M.M. Lossless Compression of Medical Images Using 3-D Predictors. *IEEE Trans. Med. Imaging* **2017**, *36*, 2250–2260. [CrossRef]
5. Manjón, J.V.; Coupé, P.; Buades, A. MRI noise estimation and denoising using non-local PCA. *Med. Image Anal.* **2015**, *22*, 35–47. [CrossRef]
6. St-Jean, S.; Coupé, P.; Descoteaux, M. Non Local Spatial and Angular Matching: Enabling higher spatial resolution diffusion MRI datasets through adaptive denoising. *Med. Image Anal.* **2016**, *32*, 115–130. [CrossRef]
7. Tashan, T.; Al-Azawi, M. Multilevel magnetic resonance imaging compression using compressive sensing. *IET Image Process.* **2018**, *12*, 2186–2191. [CrossRef]
8. Thung, K.H.; Yap, P.T.; Adeli, E.; Lee, S.W.; Shen, D. Conversion and time-to-conversion predictions of mild cognitive impairment using low-rank affinity pursuit denoising and matrix completion. *Med. Image Anal.* **2018**, *45*, 68–82. [CrossRef]
9. Chung, M.K.; Qiu, A.; Seo, S.; Vorperian, H.K. Unified heat kernel regression for diffusion, kernel smoothing and wavelets on manifolds and its application to mandible growth modeling in CT images. *Med. Image Anal.* **2015**, *22*, 63–76. [CrossRef]

10. Irrera, P.; Bloch, I.; Delplanque, M. A flexible patch based approach for combined denoising and contrast enhancement of digital X-ray images. *Med. Image Anal.* **2016**, *28*, 33–45. [CrossRef]
11. Schirrmacher, F.; Köhler, T.; Endres, J.; Lindenberger, T.; Husvogt, L.; Fujimoto, J.G.; Hornegger, J.; Dörfler, A.; Hoelter, P.; Maier, A.K. Temporal and volumetric denoising via quantile sparse image prior. *Med. Image Anal.* **2018**, *48*, 131–146. [CrossRef]
12. Xu, Z.; Gao, M.; Papadakis, G.Z.; Luna, B.; Jain, S.; Mollura, D.J.; Bagci, U. Joint solution for PET image segmentation, denoising, and partial volume correction. *Med. Image Anal.* **2018**, *46*, 229–243. [CrossRef]
13. Chen, Z.; Pazdernik, M.; Zhang, H.; Wahle, A.; Guo, Z.; Bedanova, H.; Kautzner, J.; Melenovsky, V.; Kovarnik, T.; Sonka, M. Quantitative 3D Analysis of Coronary Wall Morphology in Heart Transplant Patients: OCT-Assessed Cardiac Allograft Vasculopathy Progression. *Med. Image Anal.* **2018**, *50*, 95–105. [CrossRef]
14. Fang, L.; Li, S.; Kang, X.; Izatt, J.A.; Farsiu, S. 3-D adaptive sparsity based image compression with applications to optical coherence tomography. *IEEE Trans. Med. Imaging* **2015**, *34*, 1306–1320. [CrossRef]
15. Wu, H.; Huynh, T.T.; Souvenir, R. Echocardiogram enhancement using supervised manifold denoising. *Med. Image Anal.* **2015**, *24*, 41–51. [CrossRef]
16. Su, H.; Qi, W.; Yang, C.; Aliverti, A.; Ferrigno, G.; De Momi, E. Deep neural network approach in human-like redundancy optimization for anthropomorphic manipulators. *IEEE Access* **2019**, *7*, 124207–124216. [CrossRef]
17. Su, H.; Yang, C.; Mdeihly, H.; Rizzo, A.; Ferrigno, G.; De Momi, E. Neural network enhanced robot tool identification and calibration for bilateral teleoperation. *IEEE Access* **2019**, *7*, 122041–122051. [CrossRef]
18. Zhang, C.; Liang, T.; Mok, P.K.T.; Yu, W. FPGA Implementation of the Coupled Filtering Method and the Affine Warping Method. IEEE Trans. *Nanobioscience* **2017**, *16*, 314–325. [CrossRef]
19. Diagnostic Imaging FPGA Applications-Intel®FPGA. Available online: https://www.intel.com/content/www/us/en/healthcare-it/products/programmable/applications/diagnostic-imaging.html (accessed on 12 January 2020).
20. Medical Imaging with CT, MRI and PET. Available online: https://www.xilinx.com/applications/medical/medical-imaging-ct-mri-pet.html#overview (accessed on 12 January 2020).
21. Parikh, S.S.; Ruiz, D.; Kalva, H.; Fernandez-Escribano, G.; Adzic, V. High Bit-Depth Medical Image Compression with HEVC. *IEEE J. Biomed. Health Inform.* **2018**, *22*, 552–560. [CrossRef]
22. Pichat, J.; Iglesias, J.E.; Yousry, T.; Ourselin, S.; Modat, M. A Survey of Methods for 3D Histology Reconstruction. *Med. Image Anal.* **2018**, *46*, 73–105. [CrossRef]
23. Lahmiri, S. Comparative study of ECG signal denoising by wavelet thresholding in empirical and variational mode decomposition domains. *Healthc. Technol. Lett.* **2014**, *1*, 104–109. [CrossRef]
24. Upadhyay, J.; Mishra, B.; Patel, P. Video Denoising and Quality Improvement Using New Thresholding Based Dwt & Dammw Algorithm. In Proceedings of the 2018 IEEE International Students' Conference on Electrical, Electronics and Computer Science (SCEECS), Bhopal, India, 24–25 February 2018.
25. Song, X.; Huang, Q.; Chang, S.; He, J.; Wang, H. Three-dimensional separate descendant-based SPIHT algorithm for fast compression of high-resolution medical image sequences. *IET Image Process.* **2016**, *11*, 80–87. [CrossRef]
26. Naveen, C.; Gupta, T.V.S.; Satpute, V.R.; Gandhi, A.S. A simple and efficient approach for medical image security using chaos on EZW. In Proceedings of the 2015 Eighth International Conference on Advances in Pattern Recognition (ICAPR), Kolkata, India, 4–7 January 2015; pp. 1–6.
27. Bailey, D.G. *Design for Embedded Image Processing on FPGAs*; John Wiley & Sons (Asia) Pte Ltd.: Singapore, 2011; ISBN 9780470828519.
28. Meyer-Baese, U. Digital Signal Processing with Field Programmable Gate Arrays. In *Signals and Communication Technology*; Springer: Berlin/Heidelberg, Germany, 2007; ISBN 3-540-413413-3.
29. Madanayake, A.; Cintra, R.J.; Dimitrov, V.; Bayer, F.M.; Wahid, K.A.; Kulasekera, S.; Edirisuriya, A.; Potluri, U.S.; Madishetty, S.K.; Rajapaksha, N. Low-power VLSI architectures for DCT/DWT: Precision vs approximation for HD video, biomedical, and smart antenna applications. *IEEE Circuits Syst. Mag.* **2015**, *15*, 25–47. [CrossRef]
30. Jiang, R.M.; Crookes, D. FPGA Implementation of 3D Discrete Wavelet Transform for Real-Time Medical Imaging. In Proceedings of the 2007 18th European Conference on Circuit Theory and Design, Seville, Spain, 27–30 August 2007; pp. 519–522.

31. Ahmad, A.; Krill, B.; Amira, A.; Rabah, H. 3D Haar wavelet transform with dynamic partial reconfiguration for 3D medical image compression. In Proceedings of the 2009 IEEE Biomedical Circuits and Systems Conference, Beijing, China, 26–28 November 2009; pp. 137–140.
32. Arafa, A.A.; Saleh, H.I.; Ashour, M.; Salem, A. FFT- and DWT-Based FPGA realization of pulse shape discrimination in PET system. In Proceedings of the 2009 4th IEEE International Conference on Design and Technology of Integrated Systems in Nanoscale Era, Cairo, Egypt, 6–9 April 2009; pp. 299–302.
33. Ballesteros Larrotta, D.M.; Moreno Enciso, D.M.; Gaona Barrera, A.E. Compression of biomedical Signals on FPGA by DWT and run-length. In Proceedings of the 2010 IEEE ANDESCON, Bogota, Colombia, 15–17 September 2010; pp. 1–5.
34. Ahmad, A.; Amira, A.; Guarisco, M.; Rabah, H.; Berviller, Y. Efficient implementation of a 3-D medical imaging compression system using CAVLC. In Proceedings of the 2010 IEEE International Conference on Image Processing, Hong Kong, China, 26–29 September 2010; pp. 3773–3776.
35. Ja'Afar, N.H.; Ahmad, A.; Amira, A. Rapid prototyping of three-dimensional transform for medical image compression. In Proceedings of the 2012 11th International Conference on Information Science, Signal Processing and their Applications (ISSPA), Montreal, QC, Canada, 2–5 July 2012; pp. 842–847.
36. Li, N.Q.; Nie, Y.J.; Zhu, W. The application of FPGA-based discrete wavelet transform system in EEG analysis. In Proceedings of the 2012 International Conference on Intelligent Systems Design and Engineering Applications (ISDEA), Sanya, China, 6–7 January 2012; pp. 1306–1309.
37. Ahmad, A.; Ja'afar, N.H.; Amira, A. FPGA-based implementation of 3-D Daubechies for medical image compression. In Proceedings of the 2012 IEEE-EMBS Conference on Biomedical Engineering and Sciences, Langkawi, Malaysia, 17–19 December 2012; pp. 683–688.
38. Ja'Afar, N.H.; Ahmad, A.; Amira, A. Distributed arithmetic architecture of Discrete Wavelet Transform (DWT) with hybrid method. In Proceedings of the IEEE International Conference on Electronics, Circuits, and Systems, Abu Dhabi, United Arab Emirates, 8–11 December 2013; pp. 501–507.
39. El Hassan, E.M.; Karim, M. An FPGA-based implementation of a pre-processing stage for ECG signal analysis using DWT. In Proceedings of the 2014 Second World Conference on Complex Systems (WCCS), Agadir, Morocco, 10–12 November 2014; pp. 649–654.
40. Vijendra, V.; Kulkarni, M. ECG signal filtering using DWT haar wavelets coefficient techniques. In Proceedings of the 2016 International Conference on Emerging Trends in Engineering, Technology and Science (ICETETS), Pudukkottai, India, 24–26 February 2016; pp. 1–6.
41. Elsayed, M.; Badawy, A.; Mahmuddin, M.; Elfouly, T.; Mohamed, A.; Abualsaud, K. FPGA implementation of DWT EEG data compression for wireless body sensor networks. In Proceedings of the 2016 IEEE Conference on Wireless Sensors (ICWiSE), Langkawi, Malaysia, 10–12 October 2016; pp. 21–25.
42. Sharmila, A.; Geethanjali, P. DWT Based Detection of Epileptic Seizure From EEG Signals Using Naive Bayes and k-NN Classifiers. *IEEE Access* **2016**, *4*, 7716–7727. [CrossRef]
43. Chervyakov, N.; Lyakhov, P.; Kaplun, D.; Butusov, D.; Nagornov, N. Analysis of the quantization noise in discrete wavelet transform filters for image processing. *Electronics* **2018**, *7*, 135. [CrossRef]
44. Kim, J.; Lee, J.; Lee, S.; Lee, M. Development of 3-D stereo endoscopic PACS viewer. In Proceedings of the ISIE 2001. 2001 IEEE International Symposium on Industrial Electronics Proceedings (Cat. No.01TH8570), Pusan, Korea, 12–16 June 2001; Volume 1, pp. 278–280.
45. Lalithakumari, S.; Pandian, R.; Rani, J.; Vinothkumar, D.; Sneha, A.; Joe, B. Selection of optimum compression algorithms based on the characterization on feasibility for medical image. *Biomed. Res.* **2017**, *28*, 5633–5637.
46. Daubechies, I. *Ten Lectures on Wavelets*; Society for Industrial and Applied Mathematics: Philadelphia, PA, USA, 1992; ISBN 978-0-89871-274-2.
47. Bustin, A.; Voilliot, D.; Menini, A.; Felblinger, J.; de Chillou, C.; Burschka, D.; Bonnemains, L.; Odille, F. Isotropic Reconstruction of MR Images Using 3D Patch-Based Self-Similarity Learning. *IEEE Trans. Med. Imaging* **2018**, *37*, 1932–1942. [CrossRef]
48. Vaidyanathan, P.P. *Multirate Systems and Filter Banks*; Prentice Hall: Upper Saddle River, NJ, USA, 1993; ISBN 0136057187.
49. Rao, K.R.; Yip, P.C. *The Transform and Data Compression Handbook*; CRC Press: Boca Raton, FL, USA, 2001; ISBN 9780849336928.

50. Ravichandran, D.; Ahamad, M.G.; Dhivakar, M.R.A. Performance analysis of three-dimensional medical image compression based on discrete wavelet transform. In Proceedings of the 2016 22nd International Conference on Virtual System & Multimedia (VSMM), Kuala Lumpur, Malaysia, 17–21 October 2016; pp. 1–8.
51. Basso, A.; Cavagnino, D.; Pomponiu, V.; Vernone, A. Blind Watermarking of Color Images Using Karhunen-Loeve Transform Keying. *Comput. J.* **2011**, *54*, 1076–1090. [CrossRef]
52. Wang, Z.; Bovik, A.C.; Sheikh, H.R.; Simoncelli, E.P. Image quality assessment: From error visibility to structural similarity. *IEEE Trans. Image Process.* **2004**, *13*, 600–612. [CrossRef]
53. Seber, G.A.F.; Wild, C.J. Nonlinear Regression. In *Wiley Series in Probability and Statistics*; John Wiley & Sons, Inc.: Hoboken, NJ, USA, 1989; ISBN 9780471725312.
54. Maddala, G.S.; Lahiri, K. *Introduction to Econometrics*; Wiley: Hoboken, NJ, USA, 2009; ISBN 0470015128.
55. F-Distribution Tables. Available online: http://socr.ucla.edu/Applets.dir/F_Table.html#FTable0.001 (accessed on 12 January 2020).

© 2020 by the authors. Licensee MDPI, Basel, Switzerland. This article is an open access article distributed under the terms and conditions of the Creative Commons Attribution (CC BY) license (http://creativecommons.org/licenses/by/4.0/).

Article

Maximum Correntropy Criterion Based l_1-Iterative Wiener Filter for Sparse Channel Estimation Robust to Impulsive Noise

Junseok Lim

Department of Electrical Engineering, Sejong University, Seoul 143-747, Korea; jslim@sejong.ac.kr;
Tel.: +82-2-3408-3299

Received: 20 December 2019; Accepted: 20 January 2020; Published: 21 January 2020

Abstract: In this paper, we propose a new sparse channel estimator robust to impulsive noise environments. For this kind of estimator, the convex regularized recursive maximum correntropy (CR-RMC) algorithm has been proposed. However, this method requires information about the true sparse channel to find the regularization coefficient for the convex regularization penalty term. In addition, the CR-RMC has a numerical instability in the finite-precision cases that is linked to the inversion of the auto-covariance matrix. We propose a new method for sparse channel estimation robust to impulsive noise environments using an iterative Wiener filter. The proposed algorithm does not need information about the true sparse channel to obtain the regularization coefficient for the convex regularization penalty term. It is also numerically more robust, because it does not require the inverse of the auto-covariance matrix.

Keywords: mathematical models of digital signal processing; digital filtering; maximum correntropy; impulsive noise; sparse channel estimation

1. Introduction

In many signal processing applications [1–4], we find various sparse channels in which most of the impulse responses are close to zero and only some of them are large. In recent years, many kinds of sparse adaptive filtering algorithms have been proposed for sparse system estimation, including recursive least squares (RLS)-based [5–9] and least mean square (LMS)-based algorithms [10–14]. It is generally known that RLS-based algorithms have faster convergence and less error after convergence than LMS-based algorithms [15]. However, there are fewer RLS-based than LMS-based algorithms. Among these, the convex regularized recursive least squares (CR-RLS) proposed by Eksioglu [6] is a full recursive convex regularized RLS like a typical RLS.

While the aforementioned algorithms typically show good performance in a Gaussian noise environment, their performance deteriorates in a nonGaussian noise environment such as an impulsive noise environment. Recently, the maximum correntropy criterion (MCC) [16–19] has been successfully applied to various adaptive algorithms robust to impulsive noise. Current studies in robust sparse adaptive methods have resulted in the development of CR-RLS-based algorithms with MCC [20,21], and showed strong robustness under impulsive noise. However, CR-RLS used in [20,21] is not practical when determining the regularization coefficient for the sparse regularization term because CR-RLS [6] needs information about the true channel when calculating the regularization coefficients. In addition, MCC CR-RLS algorithms (so called convex regularized recursive maximum correntropy (CR-RMC)) [20,21] include the inversion of the auto-covariance matrix, which is linked to the numerical instability in finite-precision environments [15].

The recursive inverse (RI) algorithm [22,23] and the iterative Wiener filter (IWF) algorithm [24] have recently been proposed. RI and IWF have the same structure besides a step size calculation. They perform similarly to the conventional RLS algorithm in terms of convergence and mean squared error, without using the inverse of the auto-covariance matrix. Therefore, RI [22,23] and IWF [24] can be considered algorithms without the numerical instability of RLS.

This paper proposes a sparse channel estimation algorithm robust to impulse noise using IWF and maximum correntropy criterion with l_1-norm regularization. The proposed algorithm includes a new regularization coefficient calculation method for l_1-norm regularization that does not require information about true channels. In addition, the proposed algorithm has numerical stability because it does not include inverse matrix calculation.

In Section 2 of this paper, we derive the new algorithm using IWF. In Section 3, we provide simulation results that show the performance of the proposed algorithm. In Section 4, we note our conclusions.

2. MCC l_1-IWF Formulation

In the channel estimation problem, we assume that at time instant n the observed signal $y(n)$ is the result of the input signal $x(k)$ sequence passing through the system $\mathbf{w}_o = [w_0, \cdots, w_{M-1}]^T$ in the M-dimensional finite impulse response (FIR) format. Especially, in the sparse channel estimation problem, we assume that the system response w is sparse.

In the adaptive channel estimation, we apply an M dimensional channel $\mathbf{w}(k)$ to the same dimensional signal vector $\mathbf{x}(k)$, estimate an output $\hat{y}(k) = \mathbf{x}^T(k)\mathbf{w}(k)$, and calculate the error signal $e(k) = y(k) + n(k) - \hat{y}(k) = \widetilde{y}(k) - \hat{y}(k)$, where $y(k)$ is the output of the actual system, $\hat{y}(k)$ is the estimated output, and $n(k)$ is the measurement noise. Especially, the measurement noise is nonGaussian.

To estimate the channel in nonGaussian noise, we define an MCC cost function with exponential forgetting factor λ shown in (1) [20,21] and minimize it adaptively.

$$\underset{\hat{\mathbf{w}}(n)}{\text{minimize}} \left\{ \sum_{m=0}^{n} \lambda^{n-m} \exp\left(-\frac{\left(y(m) - \hat{\mathbf{w}}(n)^T \mathbf{x}(m)\right)^2}{2\sigma^2}\right), \text{s.t.} \|\hat{\mathbf{w}}(n)\|_1 \leq c \right\}, \quad (1)$$

where $\hat{\mathbf{w}}(n) = [\hat{w}_0(n), \cdots, \hat{w}_{M-1}(n)]^T$, $\mathbf{x}(m) = [x(m), x(m-1), \cdots, x(m-M+1)]^T$, λ is a forgetting factor, and $\|\hat{\mathbf{w}}(n)\|_1 \triangleq \sum_{k=0}^{M-1} |\hat{w}_k(n)|$. The Lagrangian for (1) becomes

$$J(\hat{\mathbf{w}}(n), \gamma(n)) = \zeta(\hat{\mathbf{w}}(n)) + \gamma(n)\left(\|\hat{\mathbf{w}}(n)\|_1 - c\right), \quad (2)$$

where $\zeta(\hat{\mathbf{w}}(n)) = \sum_{m=0}^{n} \lambda^{n-m} \exp\left(-\frac{(y(m) - \hat{\mathbf{w}}(n)^T \mathbf{x}(m))^2}{2\sigma^2}\right)$, and $\gamma(n)$ is a real-valued Lagrangian multiplier. We minimize the regularized cost function to find the optimal vector in the same way that IWF was derived [24].

The regularized cost function is convex and nondifferentiable; therefore, subgradient analysis replaces the gradient. When denoting a subgradient vector of f at $\hat{\mathbf{w}}$ with $\nabla^s f(\hat{\mathbf{w}})$, the subgradient vector of $J(\hat{\mathbf{w}}(n), \gamma(n))$ with respect to $\hat{\mathbf{w}}(n)$ can be written as follows:

$$\nabla^s J(\hat{\mathbf{w}}(n), \gamma(n)) = \nabla \zeta(\hat{\mathbf{w}}(n)) + \gamma(n) \nabla^s \left(\|\hat{\mathbf{w}}(n)\|_1\right). \quad (3)$$

Hence, for the optimal $\hat{\mathbf{w}}(n)$ minimizing $J(\hat{\mathbf{w}}(n), \gamma(n))$, we set the subgradient of $J(\hat{\mathbf{w}}(n), \gamma(n))$ to 0 at the optimal point. When evaluating the gradient $\nabla \zeta(\hat{\mathbf{w}}(n))$, we can derive a gradient vector as (4).

$$\nabla^s J(\hat{\mathbf{w}}(n), \gamma(n)) = \frac{1}{\sigma^2}(\mathbf{\Phi}(n)\hat{\mathbf{w}}(n) - \mathbf{r}(n)) + \gamma(n)\text{sgn}(\hat{\mathbf{w}}(n)) = \mathbf{g}_n, \quad (4)$$

where $e(n) = y(n) - \hat{\mathbf{w}}(n)^T \mathbf{x}(n)$, $\mathbf{\Phi}(n) = \sum_{m=0}^{n} \lambda^{n-m} \mathbf{x}(m) \mathbf{x}(m)^T = \lambda \mathbf{\Phi}(n-1) + \exp\left(-\frac{e(n)^2}{2\sigma^2}\right) \mathbf{x}(n) \mathbf{x}(n)^T$,
$\mathbf{r}(n) = \sum_{m=0}^{n} \lambda^{n-m} y(m) \mathbf{x}(m) = \lambda \mathbf{r}(n-1) + \exp\left(-\frac{e(n)^2}{2\sigma^2}\right) y(n) \mathbf{x}(n)$, and $\nabla^s(\|\hat{\mathbf{w}}(n)\|_1) = \text{sgn}(\hat{\mathbf{w}}(n))$ [6].
Using Equation (4), we can obtain the update expression for $\hat{\mathbf{w}}(n)$ as (5).

$$\hat{\mathbf{w}}(n+1) = \hat{\mathbf{w}}(n) - \mu_n \nabla^s J(\hat{\mathbf{w}}(n), \gamma(n)) = \hat{\mathbf{w}}(n) - \mu_n \mathbf{g}_n \quad (5)$$

To get the step size μ_n, we find the μ_n that minimizes exponentially averaged *a posteriori* error energy, $J(\hat{\mathbf{w}}(n+1), \gamma(n))$, where *a posteriori* error is $e(n) = y(n) - \hat{\mathbf{w}}(n+1)^T \mathbf{x}(n)$.

$$\begin{aligned}
\nabla_\mu J(\hat{\mathbf{w}}(n+1), \gamma(n)) &= -\frac{1}{\sigma^2} \hat{\mathbf{w}}(n+1)^T \mathbf{\Phi}(n) \mathbf{g}_n + \frac{1}{\sigma^2} \mathbf{r}(n)^T \mathbf{g}_n - \gamma(n) \nabla^s(\|\hat{\mathbf{w}}(n+1)\|_1)^T \mathbf{g}_n \\
&\cong -\frac{1}{\sigma^2} \hat{\mathbf{w}}(n+1)^T \mathbf{\Phi}(n) \mathbf{g}_n + \frac{1}{\sigma^2} \mathbf{r}(n)^T \mathbf{g}_n - \gamma(n) \nabla^s(\|\hat{\mathbf{w}}(n)\|_1)^T \mathbf{g}_n \\
&= -\frac{1}{\sigma^2} \hat{\mathbf{w}}(n+1)^T \mathbf{\Phi}(n) \mathbf{g}_n + \frac{1}{\sigma^2} \mathbf{r}(n)^T \mathbf{g}_n - \gamma(n) \text{sgn}(\hat{\mathbf{w}}(n))^T \mathbf{g}_n.
\end{aligned} \quad (6)$$

Substituting Equation (5) into Equation (6), we get

$$\nabla_\mu J(\hat{\mathbf{w}}(n+1), \gamma(n)) = -\frac{1}{\sigma^2} \hat{\mathbf{w}}(n)^T \mathbf{\Phi}(n) \mathbf{g}_n + \frac{1}{\sigma^2} \mu_n \mathbf{g}_n^T \mathbf{\Phi}(n) \mathbf{g}_n + \frac{1}{\sigma^2} \mathbf{r}(n)^T \mathbf{g}_n - \gamma(n) \text{sgn}(\hat{\mathbf{w}}(n))^T \mathbf{g}_n. \quad (7)$$

To find μ_n, we set $\nabla_\mu J(\hat{\mathbf{w}}(n), \gamma(n)) = 0$, and

$$\mu_n = \frac{\frac{1}{\sigma^2}(\hat{\mathbf{w}}(n)^T \mathbf{\Phi}(n) - \mathbf{r}(n)^T) \mathbf{g}_n + \gamma(n) \text{sgn}(\hat{\mathbf{w}}(n))^T \mathbf{g}_n}{\frac{1}{\sigma^2} \mathbf{g}_n^T \mathbf{\Phi}(n) \mathbf{g}_n} = \sigma^2 \frac{\mathbf{g}_n^T \mathbf{g}_n}{\mathbf{g}_n^T \mathbf{\Phi}(n) \mathbf{g}_n} \quad (8)$$

We have to derive regularization coefficient $\hat{\gamma}(n)$ such that $\|\hat{\mathbf{w}}(n+1)\|_1 = c$, i.e., the l_1-norm of vector $\hat{\mathbf{w}}(n+1)$ is preserved at all time steps of n. This can be represented by a flow equation in continuous time-domain in [25].

$$\frac{\partial \|\hat{\mathbf{w}}(t)\|_1}{\partial t} = \left(\frac{\partial \|\hat{\mathbf{w}}(t)\|_1}{\partial \hat{\mathbf{w}}}\right)^T \frac{\partial \hat{\mathbf{w}}}{\partial t} = \left(\nabla^s \|\hat{\mathbf{w}}(t)\|_1\right)^T \frac{\partial \hat{\mathbf{w}}}{\partial t} = 0. \quad (9)$$

Using a sufficiently small interval δ, the time derivative in (9) can be approximated as

$$\left(\nabla^s \|\hat{\mathbf{w}}(t)\|_1\right)^T \frac{\partial \hat{\mathbf{w}}}{\partial t} \cong \left(\nabla^s \|\hat{\mathbf{w}}(n)\|_1\right)^T \frac{(\hat{\mathbf{w}}(n+1) - \hat{\mathbf{w}}(n))}{\delta} = 0. \quad (10)$$

Using (5) and (6), (10) becomes

$$\text{sgn}(\hat{\mathbf{w}}(n))^T (\hat{\mathbf{w}}(n+1) - \hat{\mathbf{w}}(n)) = \text{sgn}(\hat{\mathbf{w}}(n))^T (-\mu_n \mathbf{g}_n) = 0. \quad (11)$$

and

$$\text{sgn}(\hat{\mathbf{w}}(n))^T \left(\frac{1}{\sigma^2} \mathbf{\Phi}(n) \mathbf{w}(n) - \frac{1}{\sigma^2} \mathbf{r}(n) + \gamma(n) \text{sgn}(\hat{\mathbf{w}}(n))\right) = 0. \quad (12)$$

The regularization coefficient $\hat{\gamma}(n)$ obtained from Equation (12) is as follows.

$$\gamma(n) = -\frac{\text{sgn}(\hat{\mathbf{w}}(n))^T (\mathbf{\Phi}(n) \mathbf{w}(n) - \mathbf{r}(n))}{\sigma^2 \text{sgn}(\hat{\mathbf{w}}(n))^T \text{sgn}(\hat{\mathbf{w}}(n))}. \quad (13)$$

On the contrary, CR-RMC algorithm in [20] uses the same regularization coefficient as that in [6]. The regularization coefficient is shown in (14)

$$\hat{\gamma}(n) = 2 \frac{\frac{tr(\Phi^{-1}(n))}{M}\left(\|\hat{\mathbf{w}}(n)\|_1 - \rho\right) + \text{sgn}(\hat{\mathbf{w}}(n))^T \Phi^{-1}(n)\varepsilon(n)}{\sigma^2 \|\Phi^{-1}(n)\text{sgn}(\hat{\mathbf{w}}(n))\|_2^2}, \quad (14)$$

where $\varepsilon(n) = \widetilde{\mathbf{w}}(n) - \hat{\mathbf{w}}(n)$ and $\widetilde{\mathbf{w}}(n)$ is the solution to the normal equation, $\Phi(n)\widetilde{\mathbf{w}}(n) = \mathbf{r}(n)$. In (14), the regularization coefficient has the parameter, ρ. In [6] and [20], the parameter was set as $\rho = f(\mathbf{w}_{true}) = \|\mathbf{w}_{true}\|_1$, with $\mathbf{w}_{true}$ indicating the impulse response of the true channel. There was no further discussion about how to set ρ. We summarize the algorithm in Table 1.

Table 1. Summary of the l_1- iterative Wiener filter (IWF).

Initialization: $\Phi(0), \mathbf{r}(0), \hat{\mathbf{w}}(0), \sigma^2$.
For n = 1 ...
$\Phi(n) = \lambda \Phi(n-1) + \exp\left(-\frac{e(n)^2}{2\sigma^2}\right)\mathbf{x}(n)\mathbf{x}(n)^T$ $\mathbf{r}(n) = \lambda \mathbf{r}(n-1) + \exp\left(-\frac{e(n)^2}{2\sigma^2}\right)y(n)\mathbf{x}(n)$ $\gamma(n) = -\frac{\text{sgn}(\hat{\mathbf{w}}(n))^T(\Phi(n)\mathbf{w}(n)-\mathbf{r}(n))}{\sigma^2 \text{sgn}(\hat{\mathbf{w}}(n))^T \text{sgn}(\hat{\mathbf{w}}(n))}$ $\mathbf{g}_n = \Phi(n)\hat{\mathbf{w}}(n) - \mathbf{r}(n) + \gamma(n)\text{sgn}(\hat{\mathbf{w}}(n))$ $\mu_n = \frac{\mathbf{g}_{n-1}^T \mathbf{g}_{n-1}}{\mathbf{g}_{n-1}^T \Phi(n) \mathbf{g}_{n-1}}$ $\hat{\mathbf{w}}(n+1) = \hat{\mathbf{w}}(n) - \mu_n \mathbf{g}_n$
end

3. Simulation Results

In this section, we compare the sparse channel estimation performance between the proposed algorithm and the convex regularized recursive maximum correntropy (CR-RMC) [20]. In addition, the numerical robustness of the proposed algorithm is compared with that of CR-RMC in the finite-precision environments.

3.1. Estimation of Sparse Channels

In this experiment, we showed the sparse system estimation results. The simulation was performed under the same experimental conditions in [6]. The true system parameter $\mathbf{w}_o$ had an order of M = 64. Out of the 64 coefficients, there were S nonzero coefficients. The nonzero coefficients were placed randomly, and the values of the coefficients were drawn from a $N(0, 1/S)$ distribution. The impulsive noise is generated according to the Gaussian mixture model [26]

$$p_v = (1 - p_r)N(0, \sigma_1^2) + p_r N(0, \sigma_2^2) \quad (15)$$

where $N(0, \sigma_i^2) (i = 1, 2)$ denote the Gaussian distribution with zero-mean and variance σ_i^2. The p_r denotes the occurrence probability of the Gaussian distribution with variance σ_2^2, which usually is much larger than σ_1^2 so as to generate the impulsive noises. The zero-mean Gaussian distribution with variance σ_1^2 generated the background noise, and the zero-mean Gaussian distribution with variance σ_2^2 (usually $\sigma_2^2 \gg \sigma_1^2$) generated the impulsive noise with the probability p_r. In this experiment, we set the variance of σ_1^2 to 0.01 and generate the input signal so that SNR keeps 20dB. The other parameters were set as $\sigma_2^2 = 500$ and $p_r = 0.01$.

We compare CR-RMC [20] using the true system response information and the proposed algorithm using a regularization factor that did not use the true system response. It also included the results of the MCC-RLS [27] and the conventional RLS without considering impulsive noise and sparsity. For the performance evaluation, we simulated the algorithms in the sparse impulse response for S = 4, 8, 16, 32.

Figure 1 illustrates the mean standard deviation (MSD) curves. The results show that the estimation performance of the proposed algorithm is similar to that of CR-RMC using the regularization factor referring to the true system impulse response. As expected, the conventional RLS produced the worst MSD in all cases.

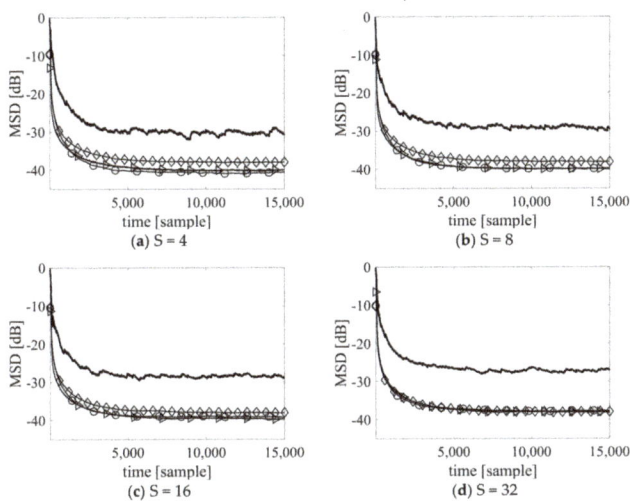

Figure 1. Steady state MSD for S = 4, 8, 16, 32 (-▷-: the proposed algorithm, -○-: convex regularized recursive maximum correntropy (CR-RMC), -◇-: maximum correntropy criterion (MCC)- recursive least squares (RLS), solid line: conventional RLS without considering impulsive noise and sparsity): (**a**) S = 4, (**b**) S = 8, (**c**) S = 16, (**d**) S = 32.

Figure 1 confirms that, without *a priori* information about the true system impulse response, the proposed regularization factor works similarly to that of the regularization factor in CR-RMC using the true system impulse response information.

3.2. Numerical Robustness Experiment

In this experiment, we showed the proposed algorithm to be numerically more robust than CR-RMC in the finite-precision environments. We performed channel estimation with finite precision by quantization to show the numerical robustness [28]. The round-off error from the quantization with finite bits was accumulated and propagated through the inverse matrix operation of $\Phi(n)$, and, finally, explosive divergence occurred [15,28]. To illustrate this, we repeated numerical stability experiments, decreasing the quantization bit from 32 bits by 1 bit to find the quantization bits that started numerical instability in each algorithm while comparing and verifying the performance for the case of S = 4 and S = 16. In addition, the rest of the setup for the experiment was the same as Experiment 3.1.

Figure 2 shows the results of comparing the performance of the proposed algorithm and CR-RMC in terms of MSD with different numbers of quantization bits. Figure 2a,b shows the results when quantized to 32 bits. In this case, we can observe that the proposed algorithm as well as CR-RMC converges normally as Figure 1a,b. Figure 2 shows the results of comparing the performance of the proposed algorithm and CR-RMC in terms of MSD with different numbers of quantization bits. Figure 2a,b shows the results when quantized to 32 bits. In this case, we can observe that the proposed algorithm as well as CR-RMC converges normally as Figure 1a,b. Figure 2c,d shows the quantization results for 16 bits. In 16 bits, CR-RMC started numerical instability. Compared with the results of Figure 2a,b, it can be observed that quantized CR-RMC diverges due to the cumulative effect of the error of quantization noise. Figure 2e,f shows the quantization results for 11 bits. In 11 bits, the proposed algorithm also started numerical instability. If we consider the level of quantization error with signal to quantization noise ratio (SQNR), SQNR (dB) = 1.76 + 6.02 × bits [29], CR-RMC is stable at above 98.08 dB in SQNR. The proposed algorithm is stable at above 67.98 dB in SQNR. In other word, the proposed algorithm has 30.1 dB gain in numerical stability compared to CR-RMC.

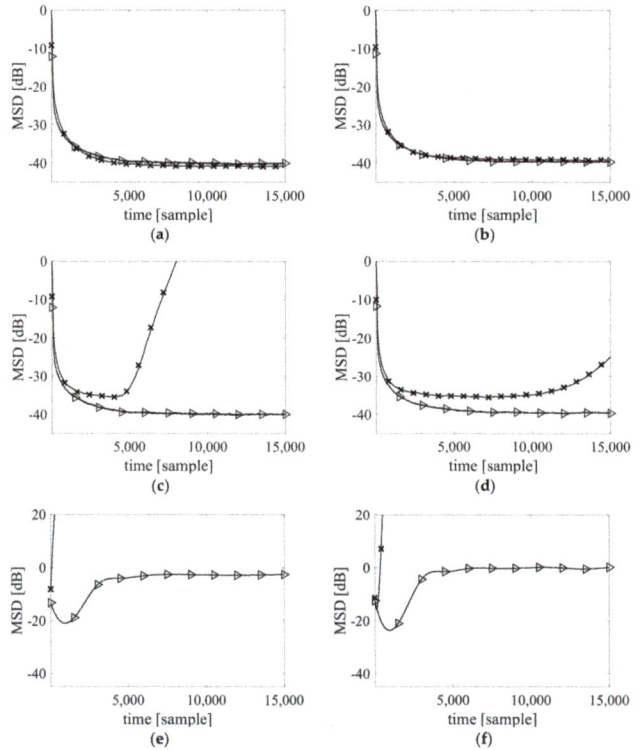

Figure 2. Results of numerical robustness experiment (-▷-: the proposed algorithm, -x-: CR-RMC): (**a**) S = 4 case quantized by 32 bits; (**b**) S = 16 case quantized by 32 bits; (**c**) S = 4 case quantized by 16 bits; (**d**) S = 16 case quantized by 16; (**e**) S = 4 case quantized by 11 bits; and (**f**) S = 16 case quantized by 11 bits.

The experimental results confirm that the proposed algorithm is numerically more robust than CR-RMC.

4. Conclusions

In this paper, this paper have proposed a sparse channel estimation algorithm robust to impulse noise using IWF and MCC with l_1-norm regularization. The proposed algorithm includes a regularization factor calculation algorithm without any requirement for *a priori* knowledge about the true system response. The simulation results show that the proposed algorithm works similarly to the CR-RMC algorithm with a regularization factor referring to the true system response information. In addition, simulation results show that the proposed algorithm is more robust against numerical error than the CR-RMC algorithm.

Funding: This research received no external funding.

Acknowledgments: This paper was supported by the Agency for Defense Development (ADD) in Korea (UD190005DD).

Conflicts of Interest: The authors declare no conflict of interest.

References

1. Loganathan, P.; Khong, A.W.; Naylor, P.A. A class of sparseness-controlled algorithms for echo cancellation. *IEEE Trans. Audio Speech Lang. Process.* **2003**, *17*, 1591–1601. [CrossRef]
2. Carbonelli, C.; Vedantam, S.; Mitra, U. Sparse channel estimation with zero tap detection. *IEEE Trans. Wirel. Commun.* **2007**, *6*, 1743–1763. [CrossRef]
3. Yousef, N.R.; Sayed, A.H.; Khajehnouri, N. Detection of fading overlapping multipath components. *Signal Process.* **2006**, *86*, 2407–2425. [CrossRef]
4. Singer, A.C.; Nelson, J.K.; Kozat, S.S. Signal processing for underwater acoustic communications. *IEEE Commun. Mag.* **2009**, *47*, 90–96. [CrossRef]
5. Babadi, B.; Kalouptsidis, N.; Tarokh, V. SPARLS: The sparse RLS algorithm. *IEEE Trans. Signal Process.* **2010**, *58*, 4013–4025. [CrossRef]
6. Eksioglu, E.M. RLS algorithm with convex regularization. *IEEE Signal Process. Lett.* **2011**, *18*, 470–473. [CrossRef]
7. Eksioglu, E.M. Sparsity regularised recursive least squares adaptive filtering. *IET Signal Process.* **2011**, *5*, 480–487. [CrossRef]
8. Das, B.; Chakraborty, M. Improved l_0-RLS adaptive filter algorithms. *Electron. Lett.* **2017**, *53*, 1650–1651. [CrossRef]
9. Liu, L.; Zhang, Y.; Sun, D. VFF l_1-norm penalised WL-RLS algorithm using DCD iterations for underwater acoustic communication. *IET Commun.* **2017**, *11*, 615–621. [CrossRef]
10. Li, Y.; Wang, Y.; Jiang, T. Norm-adaption penalized least mean square/fourth algorithm for sparse channel estimation. *Signal Process.* **2016**, *128*, 243–251. [CrossRef]
11. Jahromi, M.N.S.; Salman, M.S.; Hocanin, A. Convergence analysis of the zero-attracting variable step-size LMS algorithm for sparse system identification. *Signal Image Video Process.* **2013**, *9*, 1353–1356. [CrossRef]
12. Li, Y.; Wang, Y.; Jiang, T. Sparse-aware setmembership NLMS algorithms and their application for sparse channel estimation and echo cancelation. *AEU-Int. J. Electron. Commun.* **2016**, *70*, 895–902. [CrossRef]
13. Gu, Y.; Jin, J.; Mei, S. l_0-norm constraint LMS algorithm for sparse system identification. *IEEE Signal Process. Lett.* **2009**, *16*, 774–777.
14. Chen, Y.; Gu, Y.; Hero, A.O. Sparse LMS for system identification. In Proceedings of the IEEE International Conference on Acoustics, Speech Signal Processing, Taipei, Taiwan, 19–24 April 2009; pp. 3125–3128.
15. Haykin, S. *Adaptive Filter Theory*, 5th ed.; Prentice Hall: Upper Saddle River, NJ, USA, 2014.
16. Liu, W.; Pokharel, P.P.; Principe, J.C. Correntropy: Properties and applications in non-Gaussian signal processing. *IEEE Trans Signal Process.* **2007**, *55*, 5286–5298.
17. Chen, B.; Xing, L.; Liang, J.; Zheng, N.; Principe, J.C. Steady-state mean-square error analysis for adaptive filtering under the maximum correntropy criterion. *IEEE Signal Process. Lett.* **2014**, *21*, 880–884.
18. Ma, W.; Qu, H.; Gui, G.; Xu, L.; Zhao, J.; Chen, B. Maximum correntropy criterion based sparse adaptive filtering algorithms for robust channel estimation under non-Gaussian environments. *J. Frankl. Inst.* **2015**, *352*, 2708–2727. [CrossRef]
19. Chen, B.; Xing, L.; Zhao, H.; Zheng, N.; Principe, J.C. Generalized correntropy for robust adaptive filtering. *IEEE Trans. Signal Process.* **2016**, *64*, 3376–3387. [CrossRef]
20. Zhang, X.; Li, K.; Wu, Z.; Fu, Y.; Zhao, H.; Chen, B. Convex regularized recursive maximum correntropy algorithm. *Signal Process.* **2016**, *129*, 12–16. [CrossRef]
21. Ma, W.; Duan, J.; Chen, B.; Gui, G.; Man, W. Recursive generalized maximum correntropy criterion algorithm with sparse penalty constraints for system identification. *Asian J. Control* **2017**, *19*, 1164–1172. [CrossRef]
22. Ahmad, M.S.; Kukrer, O.; Hocanin, A. Recursive inverse adaptive filtering algorithm. *Digit. Signal Process.* **2011**, *21*, 491–496. [CrossRef]
23. Salman, M.S.; Kukrer, O.; Hocanin, A. Recursive inverse algorithm: Mean-square-error analysis. *Digit. Signal Process.* **2017**, *66*, 10–17. [CrossRef]

24. Xi, B.; Liu, Y. Iterative Wiener Filter. *Electron. Lett.* **2013**, *49*, 343–344. [CrossRef]
25. Khalid, S.; Abrar, S. Blind adaptive algorithm for sparse channel equalization using projections onto l_p-ball. *Electron. Lett.* **2015**, *51*, 1422–1424. [CrossRef]
26. Wang, W.; Zhao, J.; Qu, H.; Chen, B. A correntropy inspired variable step-size sign algorithm against impulsive noises. *Signal Process.* **2017**, *141*, 168–175. [CrossRef]
27. Ma, W.; Qu, H.; Zhao, J. Estimator with forgetting factor of correntropy and recursive algorithm for traffic network prediction. In Proceedings of the 2013 25th Chinese Control and Decision Conference (CCDC), Guiyang, China, 25–27 May 2013; pp. 490–494.
28. Sayed, A.H. *Fundamentals of Adaptive Filtering*; John Wiley & Sons, Inc.: Hoboken, NJ, USA, 2003; pp. 775–803.
29. Proakis, J.G.; Manolakis, D.K. *Digital Signal Processing*, 4th ed.; Pearson Education Limited: London, UK, 2014; p. 35.

© 2020 by the author. Licensee MDPI, Basel, Switzerland. This article is an open access article distributed under the terms and conditions of the Creative Commons Attribution (CC BY) license (http://creativecommons.org/licenses/by/4.0/).

Article

Development of Classification Algorithms for the Detection of Postures Using Non-Marker-Based Motion Capture Systems

Tatiana Klishkovskaia [1,*], Andrey Aksenov [1,2], Aleksandr Sinitca [3], Anna Zamansky [4], Oleg A. Markelov [5] and Dmitry Kaplun [3]

1. Department of Bioengineering Systems, Saint Petersburg Electrotechnical University "LETI", 197376 Saint Petersburg, Russia; a.aksenov@hotmail.com
2. Russian Ilizarov Scientific Center for Restorative Traumatology and Orthopaedics, 640014 Kurgan, Russia
3. Department of Automation and Control Processes, Saint Petersburg Electrotechnical University "LETI", 197376 Saint Petersburg, Russia; amsinitca@etu.ru (A.S.); dikaplun@etu.ru (D.K.)
4. Information Systems Department, University of Haifa, Haifa 3498838, Israel; annazam@is.haifa.ac.il
5. Centre for Digital Telecommunication Technologies, Saint Petersburg Electrotechnical University "LETI", 5 Professor Popov Street, 197376 Saint Petersburg, Russia; OAMarkelov@etu.ru
* Correspondence: taklishkovskaya@stud.eltech.ru; Tel.: +79-21-655-04-70

Received: 14 May 2020; Accepted: 7 June 2020; Published: 10 June 2020

Abstract: The rapid development of algorithms for skeletal postural detection with relatively inexpensive contactless systems and cameras opens up the possibility of monitoring and assessing the health and wellbeing of humans. However, the evaluation and confirmation of posture classifications are still needed. The purpose of this study was therefore to develop a simple algorithm for the automatic classification of human posture detection. The most affordable solution for this project was through using a Kinect V2, enabling the identification of 25 joints, so as to record movements and postures for data analysis. A total of 10 subjects volunteered for this study. Three algorithms were developed for the classification of different postures in Matlab. These were based on a total error of vector lengths, a total error of angles, multiplication of these two parameters and the simultaneous analysis of the first and second parameters. A base of 13 exercises was then created to test the recognition of postures by the algorithm and analyze subject performance. The best results for posture classification were shown by the second algorithm, with an accuracy of 94.9%. The average degree of correctness of the exercises among the 10 participants was 94.2% (SD1.8%). It was shown that the proposed algorithms provide the same accuracy as that obtained from machine learning-based algorithms and algorithms with neural networks, but have less computational complexity and do not need resources for training. The algorithms developed and evaluated in this study have demonstrated a reasonable level of accuracy, and could potentially form the basis for developing a low-cost system for the remote monitoring of humans.

Keywords: posture classification; skeleton detection; motion capture; exercise classification; virtual rehabilitation

1. Introduction

Demographic ageing in humans means that to date, 12% of the global population are aged over 60 years, and this number is likely to double within a few decades [1]. Ageing leads to a higher prevalence of complications that may benefit from exercise therapy. Such an increase in ageing will mean that the rapid development of science and medicine, as well as the introduction of new technologies and methodologies utilized by health systems, will be needed. Increased knowledge has been gained regarding new treatment regimes for a growing number of chronic diseases and traumas, but with consequential increases in social and economic costs [2]. It is well-known that rehabilitation forms an important part of a typical overall treatment plan, which can be delivered, for instance, by utilizing therapeutic exercise (physiotherapy). The performance of physical activity has many advantages in older people with dementia, and can positively affect the preservation of cognitive abilities [3]. Stroke patents may also benefit from physical activities, which can result in improved recovery rates.

However, the success of rehabilitation largely depends on keeping the patient interested and motivated in the continuation of treatment. Factors influencing adherence to the continuation of physical education depend on whether people continue to receive professional assistance and counselling after the completion of the initial training [4]. Among the main reasons for the termination of continued professional assistance and counselling are forgetfulness, a lack of further supervision and motivation, and time restraints (for example: attending the rehabilitation center).

The use of exercise therapy delivered remotely using posture recognition and interactive content may have a positive impact on enabling patients to perform exercise, as well as their willingness to continue training and rehabilitation programs [5].

Events such as the recent Covid-19 pandemic reinforce the need for remote exercise therapy with feedback from a doctor, which would be very beneficial for many patients with different disabilities.

Traditionally, exercise therapy consists of demonstrating exercises, observation and evaluation by a health professional, which in turn requires special training and significant face-to-face contact with a patient. However, modern computer and sensor technologies could be utilized to augment (or where appropriate, replace) direct intervention by health professionals. Such technologies that can capture specific postures will be able to determine whether or not the exercise regimes provided to the patient are proving the beneficial postural changes over time, with reference to those obtained from healthy adults. With the capabilities of motion capture systems advancing significantly in recent years, and with motion capture systems being more accessible and effective, they allow the kinematics of the human body to be measured and recorded with sufficient accuracy in real time, even using web cameras.

Two main types of motion capture systems are widely used: those which use markers, and those which estimate joint and limb segment parameters based on neural network training from marker systems. The first requires use of a special suit, or a removable system of sensors (active or passive markers) attached to the human body. The second type, such as those provided by Microsoft Kinect, Intel RealSense, Structure Core and others, use color and depth data, as well as image recognition algorithms, to retrieve the data. These systems can record kinematic data and perform analysis of the human body's movements in real time.

In addition, the development and availability of these sensors opens more opportunities, as it makes it possible to create bespoke courses of rehabilitation, and to monitor their implementation [6–11]. Similar applications have been developed for different patient groups, but the most widely represented software has been designed for post-stroke patients [12–16]. Software has also been designed for people with neurological diseases [17], including cerebral palsy [18], multiple sclerosis [19] and traumatic brain injuries [20].

However, the algorithms used by these systems to estimate the accuracy of execution of movements by such patients are not fully described in the literature. Two of those algorithms can, however, be distinguished by their differing mode of operation. The first is based on the use of dynamic time warping (DTW), along with fuzzy logic [7], and the other is based on the recognition of different body segment postures and trajectories [21]. However, the use of a home-based system, using virtual rehabilitation and offering the possibility of communication with a doctor, is more convenient for the patient, and also allows the course of rehabilitation to be altered by adding new exercises, if necessary. DTW is, however, difficult to apply when compared to posture estimation algorithms. Anton et al.

utilized the recognition of postures together with trajectories, which resulted in an accuracy of posture estimation of 91.9%, and detection of movements of 95.16% [21].

Recent advances in machine learning have led to the use of machine learning algorithms in many studies, including posture classification [22,23]. The objective of these studies is to classify the sitting postures via conventional algorithms and deep learning-based algorithms using the body pressure distribution data from pressure sensors [22]. After classifying the sitting postures using several classifiers, average and maximum classification rates of 97.20% and 97.94%, respectively, were obtained from nine subjects with a support vector machine using the radial basis function kernel. Through a comparison of the application of the convolutional neural network (CNN) and conventional machine learning algorithms, the effectiveness of an approach [23] wherein the CNN algorithm is applied was shown (average value of accuracy = 0.953). However, machine learning-based algorithms have problems with a computational complexity that lead to an inability of real-time implementation (in reference [22], the authors stressed this point) and the need for resources for training.

These examples of previous research in the use of posture recognition algorithms provide strong arguments for the continued research and development of such algorithms.

The aim of this research was to develop simpler and more efficient identification algorithms for posture and exercise classification within healthy participants, as well as to evaluate these using Kinect V2. The main contributions of our work can be summarized as follows. Three algorithms for the classification of different postures were developed and evaluated. The effectiveness of these algorithms was based on a total error of vector lengths and a total error of angles, and the multiplication of these two parameters was proved. To compare the effectiveness of classification algorithms, a database was created from the descriptions of the 573 known postures, as well as 903 postures which were not related to them. It was shown that the algorithms presented in this study were demonstrated to be reasonably accurate, and could potentially form the basis for developing a simple system for the remote monitoring of rehabilitation involving exercise therapy.

The remainder of this paper is organized as follows. In Section 2, we describe the Microsoft Kinect V2-based approach to the automatic classification of human exercise movement and present three algorithms for posture classifications. In Section 3, we compare the effectiveness of the three developed classification algorithms by means of a database that was created from the descriptions of the 573 known postures and 903 postures which were not correctly performed. In Section 4, we discuss the results and how they can be interpreted from the perspective of previous studies, and of the working hypotheses. Future research directions also are highlighted. Finally, we present the conclusions in Section 5.

2. Materials and Methods

2.1. Participants

Ten healthy young adults (mean ± standard deviation age: 23.4 ± 4.1 years; six males with body mass: 72.7 ± 4.7 kg and height: 179.7 ± 4.2 cm; four females with body mass: 51.5 ± 2.6 kg and height: 163.3 ± 2.8 cm) participated in forming the exercise database. A healthy male (age 35, weight 75 kg and height 184 cm) and a healthy female (age 23, weight 50 kg and height 165 cm) were used to form the independent reference posture database. This research was completed as part of the state project of the Ministry of Health of Russia and was approved by the Ethics Committee of the Ilizarov Scientific Center for Restorative Traumatology and Orthopaedics (17 May 2018, protocol No.2(57)). All participants read the information sheet before the experiment. Written informed consent was obtained from all the participants.

2.2. Posture Description

A 3D Sensor (Microsoft Kinect V2) was used to record movement, as it is able to recognize different subjects, track their movement and create a skeleton comprising 25 points (Figure 1), which may be described by three-dimensional coordinates (i.e., by using X, Y and Z planes of motion).

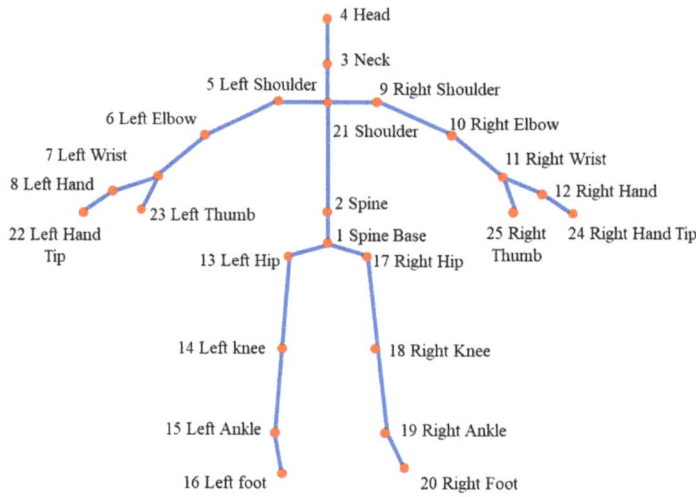

Figure 1. Diagram of connection of points received from the sensor.

Any movement consists of a series of postures. Eighteen joints were used to describe a posture in a series of volunteer subjects. It was decided to exclude joints such as those numbered 16, 20, 21, 22, 23, 24 and 25 (Figure 1) from algorithms, as they demonstrated high inconsistency in tracking accuracy. A total of 40 parameters were therefore calculated, based on 18 points: 17 were vector lengths (Table 1) and 23 were angles. However, each algorithm used a different number of parameters, as described in Section 2.3.

Table 1. Vector lengths used for the algorithm, where numbers represent the joint as shown in Figure 1.

No.	Vector Length	No.	Vector Length	No.	Vector Length
1	2-1	7	2-8	13	2-14
2	2-3	8	2-9	14	2-15
3	2-4	9	2-10	15	2-17
4	2-5	10	2-11	16	2-18
5	2-6	11	2-12	17	2-19
6	2-7	12	2-13		

The vector lengths were calculated relative to a position on the centerline of the torso (see point "2", Figure 1), as it had minimal errors in tracking. As each subject had a different body shape, this meant lengths between joints were not consistent, and it was therefore decided to normalize them using the participants' heights using the following formula [24]

$$D_{vector} = \sqrt{\frac{(x-x_0)^2 + (y-y_0)^2 + (z-z_0)^2}{height}}, \quad (1)$$

where x_0, y_0 and z_0 represent coordinates of the midpoint of the back, and x, y, z are the coordinates of the point for which the distance is calculated.

Eleven angles were used in algorithms to describe postures and movements, as shown in Figure 2 and Table 2. For all 11 joints, the angles were between two vectors in 3D space. However, for the shoulder, hip and knee, the angles were calculated in the frontal and sagittal planes only.

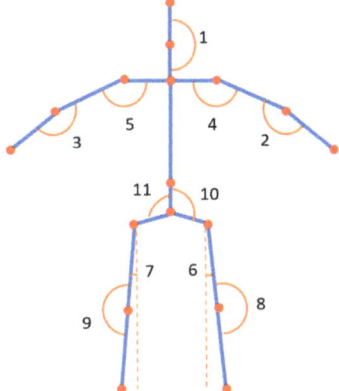

Figure 2. Angles used in describing poses.

Table 2. Angles used to describe postures.

No.	Angle	Vector Directions by Points
1	Neck tilt	[4 3] [21 3]
2	Right elbow	[9 10] [11 10]
3	Left elbow	[5 6] [7 6]
4	Right shoulder	[2 21] [10 9]
5	Left shoulder	[2 21] [6 5]
6	Right thigh	[1 2] [18 17]
7	Left thigh	[1 2] [14 13]
8	Right knee	[17 18] [19 18]
9	Left knee	[13 14] [16 14]
10	Inclination of the back to the right thigh	[2 1] [17 1]
11	Inclination of the back to the left thigh	[2 1] [13 1]

The angles were calculated as the angle between two 3D vectors

$$D_{angle} = \arccos\left(\frac{x_1 x_2 + y_1 y_2 + z_1 z_2}{\sqrt{x_1^2 + y_1^2 + z_1^2}\sqrt{x_2^2 + y_2^2 + z_2^2}}\right), \quad (2)$$

where x_n, y_n and z_n are the coordinates of vectors obtained by the differences between points, according to Table 1.

2.3. Experemental Protocol

A database of 12 postures was created to validate the algorithms containing postures and exercise movements by ten subjects (Table 3, Figures 3 and 4). Each subject was asked to do 13 exercises and repeat each one at least 25 times. Subjects were allowed to rest if they felt fatigued. On average, it took around four hours to record 13 exercise movements for each participant. Exercise movements were randomized for each subject.

Table 3. Reference database of postures for the two people recorded and used for the classification of other participants.

	Posture
1	Hand outstretched
2	Hands down (neutral posture)
3	Hands on waist
4	Right hand up
5	Left hand up
6	Both hands up
7	Hands forward
8	Right knee up (hands on waist)
9	Left knee up (hands on waist)
10	Both hands to the head
11	Right hand to the side
12	Left hand to the side

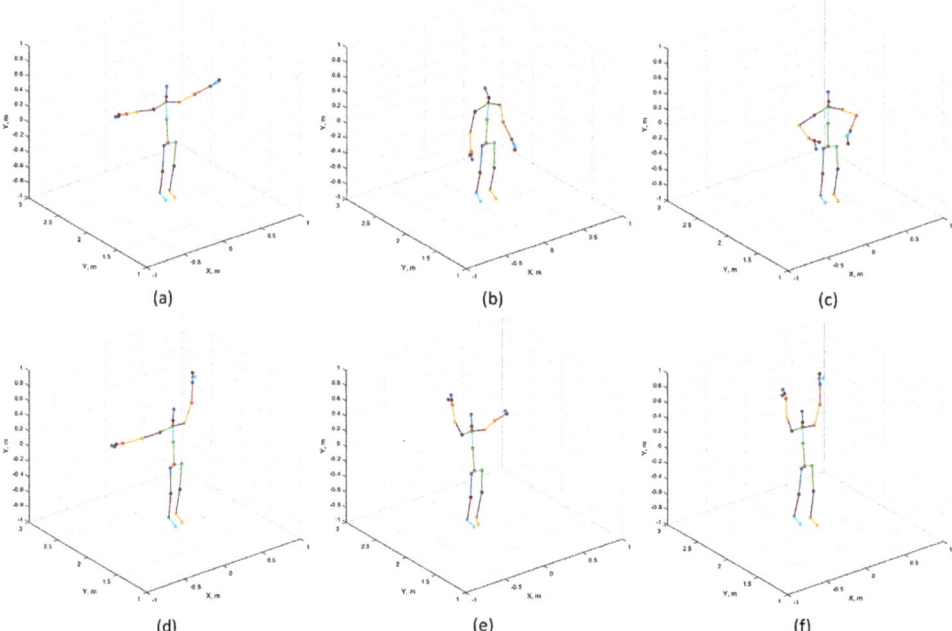

Figure 3. Postures: (**a**) hands outstretched; (**b**) hands down; (**c**) hands on the waist; (**d**) left hand up; (**e**) right hand up; and (**f**) both hands up.

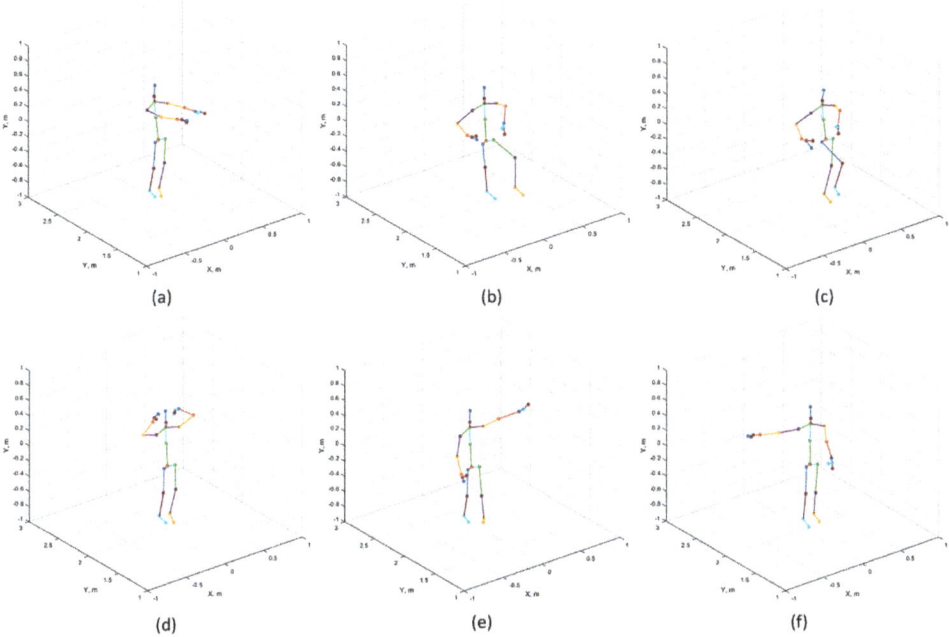

Figure 4. Postures: (**a**) hands forward; (**b**) left knee up; (**c**) right knee up; (**d**) both hands to the head; (**e**) left hand to the side; and (**f**) right hand to the side.

The movement exercises were described as a sequence of postures. The simplest movement was described by the start and the end position. In some cases, however, there were more complex sequences of movements where the middle phase movement comprised a combination of several postures. A total of thirteen different exercise test movements were eventually used in the study, as shown in Table 4.

Table 4. Test exercises.

No.	Posture Exercises (Initial Posture–Final Posture)
1	Hands down–hands outstretched
2	Hands down–hands up
3	Hands at the sides–right hand up
4	Hands at the sides–left hand up
5	Hands at the sides–hands to the head
6	Hands on the belt–right knee up
7	Hands on the belt–left knee up
8	Hands at the sides–hands forward
9	Hands down–hands forward
10	Hands up–hands forward
11	Hands forward–right hand to the side
12	Hands forward–left hand to the side
13	Hands down–hands forward–hands up–hands outstretched

2.4. Accuracy Evaluation of Postures and Movement Exercises

The accuracy, specificity and sensitivity were calculated based on formulas described in the article [25].

The classification of postures was made by comparing the recorded posture descriptors (D_i) with a reference database (D_j). The distance Er_i for each pose i between the reference and reordered posture could be calculated as:

$$Er_i = dist(D_i, D_j), \qquad (3)$$

A descriptor is composed of two parameters (angles and vectors), and thus two types of errors were calculated: the total error of the length of vectors and the total error of angles.

The first was calculated using absolute differences between them

$$ErVec_i = \sum_{k=1}^{17} |D_i(k) - D_j(k)|, \qquad (4)$$

where $D_i(k)$, k = between 1 and 17—parameters that are responsible for the length of the vectors. The total error angles for postures i were calculated using the formula

$$ErAngle_i = \sum_{k=18}^{40} |D_i(k) - D_j(k)|, \qquad (5)$$

where $D_i(k)$, k = between 18 and 40—parameters responsible for the values of angles.

Based on those types of errors, three algorithms for the posture classifications assessment were developed. To classify the posture, the results should be equal to or almost equal to the reference database, so that the algorithm can define the correct posture classification from the data set collected. This was achieved by setting a threshold for the three algorithms:

- Algorithm 1: vector length error (A1)
- Algorithm 2: angle error (A2)
- Algorithm 3: multiplication of angle errors by vector errors (A3)

To evaluate the most accurate algorithm for posture detection, the classification database was made using the descriptions of either "correct" or "incorrect" postures. In our study, all subjects were young and healthy, therefore it was enough to use two people for the posture reference database. However, the reference database would be more complex if participants had some disabilities and varied in age group.

To justify the accuracy of exercise movement classification, the database, with a set of sequenced postures in the correct order, was made, as shown in the examples in Figure 5.

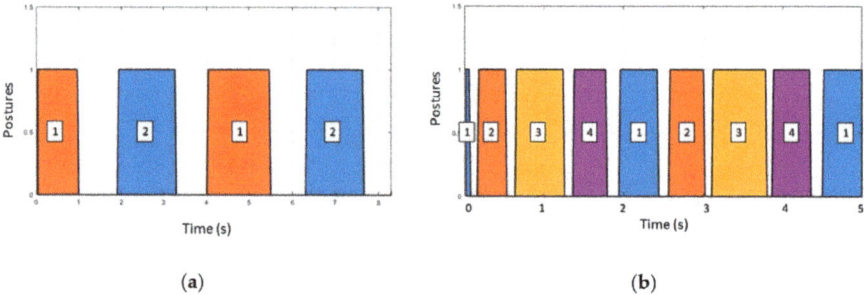

Figure 5. Example of a movement exercise: (**a**) combination of two postures; and (**b**) a more complex movement exercise with a set of postures in sequential order.

Matlab was used for data collection, analysis.

3. Results

3.1. Classification Algorithms

To compare the effectiveness of different classification algorithms, a database was created from the descriptions of the 573 known postures, as shown in Table 3, and 903 postures which were not correct. Using this database, three algorithms were obtained that tested the sensitivity, specificity and accuracy of values. (Figures 6 and 7).

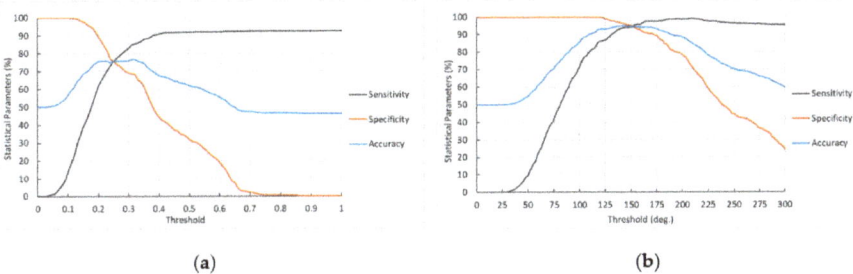

Figure 6. Relationship between specificity, sensitivity, accuracy and threshold for: (**a**) Algorithm 1; and (**b**) Algorithm 2.

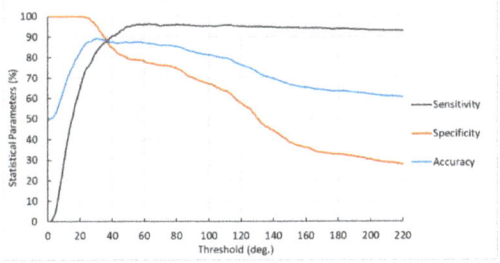

Figure 7. Relationship between specificity, sensitivity, accuracy and threshold for Algorithm 3.

The mean sensitivity for the first algorithm was 92.5%, while for the second it was 98.95% and for the third it was 96.5%. Table 5 demonstrates detailed statistical results for three algorithms. Figure 8 shows receiver operator characteristic (ROC) curve results for three algorithms.

Table 5. Statistical results.

Algorithm	Mean Sensitivity, %	Intersection of Sensitivity and Specificity, %	Mean Accuracy, %	Area under the ROC Curve
Total vector error (A1)	92.5	75.7	76.6	0.862
Total angle error (A2)	98.95	94.1	94.9	0.986
Multiplication of vector errors by angle errors (A3)	96.5	87.7	89.3	0.966

The mean intersection of sensitivity and specificity for the first algorithm was 75.7%, while for the second it was 94.1% and for the third it was 87.7%. The mean accuracy for the first algorithm was 76.6%, while for the second it was 94.9% and for the third it was 89.3%. The area under the ROC curves for the first algorithm was 0.862, while for the second it was 0.986 and for the third it was 0.966.

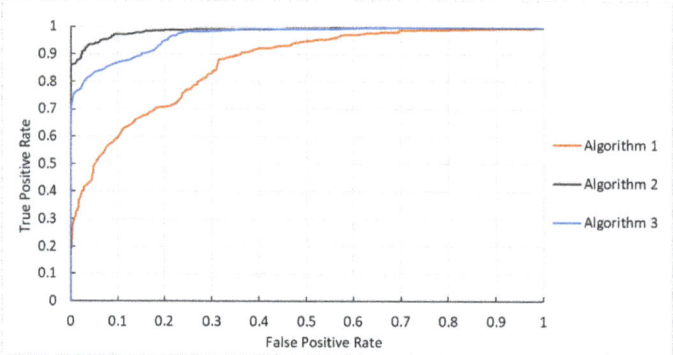

Figure 8. Relationship between false and true positive rates between three different algorithms.

3.2. Number of Exercises Performed by Participants

Each participant performed at least 390 exercises in total. Table 6 demonstrates detailed information on the number of exercises performed by each participant.

Table 6. Number of exercises performed by each participant.

	No.	Exercise Number													Total
		1	2	3	4	5	6	7	8	9	10	11	12	13	
Participants	1	25	25	35	35	25	40	40	25	25	25	35	35	40	410
	2	35	25	25	25	25	40	40	25	35	25	35	35	40	410
	3	25	25	25	25	25	40	40	25	25	25	35	35	40	390
	4	35	25	25	25	25	40	40	35	25	35	35	35	40	420
	5	25	25	25	25	25	40	40	25	25	25	35	35	40	390
	6	25	25	25	25	25	40	40	25	25	25	35	35	40	390
	7	25	25	35	35	25	40	40	25	25	35	35	35	40	420
	8	40	25	35	35	25	40	40	25	25	25	35	35	40	425
	9	25	25	35	35	25	40	40	40	25	25	35	35	40	425
	10	25	35	25	25	25	35	35	25	25	40	35	35	40	405

The highest values of accuracy for movement exercises was demonstrated by the second algorithm, with 94.3% (SD 1.7%), as shown in Figure 9.

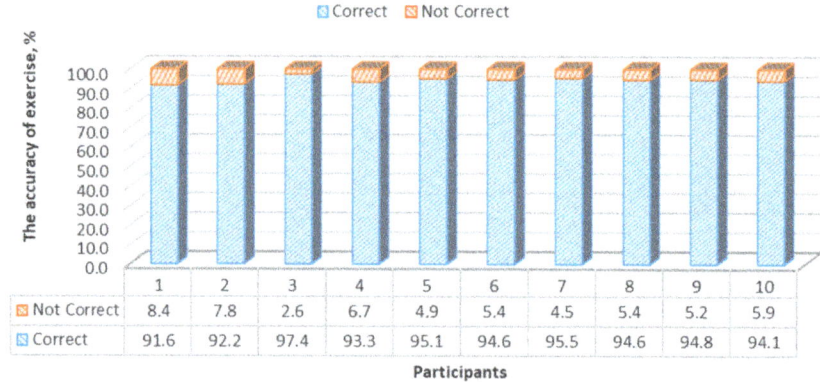

Figure 9. The accuracy of exercise movements performed by ten subjects for the second algorithm.

Figure 10 shows the percentage ratio of the identification of 13 exercises.

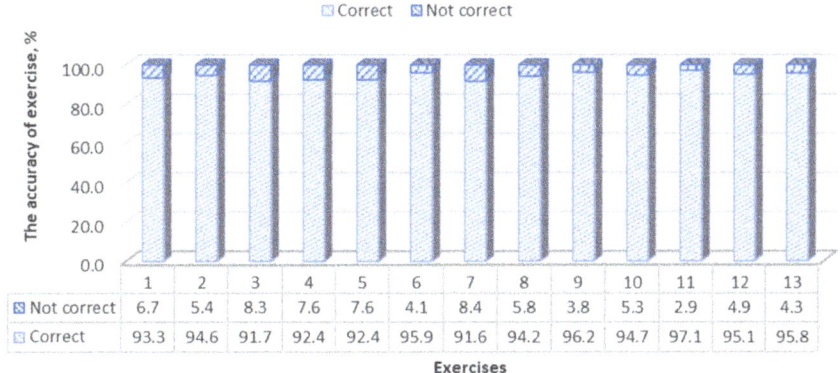

Figure 10. The identification ratio for 13 exercises with the implementation of Algorithm 2.

The average identification ratio of correct movement classification among participants was 94.3% (SD 1.7%). The average identification of correct exercises was 94.2% (SD 1.8%).

4. Discussion

The aim of this study was to determine accurate posture and exercise classification algorithms with low-cost sensors such as Microsoft Kinect, which has also led to the development of different virtual rehabilitation programs [13,26]. The use of such sensors can have many advantages. Firstly, they highlight interactivity and motivation, and they can also be used at home. This is important for people who live in remote areas, where there may not be experts who are locally available. In addition, the technique can be adapted to the needs of any patient group [27], or animals [28–31].

The comparison of this sensor with a professional optical motion capture system has demonstrated that it has the accuracy sufficient for both the tasks and data generation capability needed by specialists in the field of rehabilitation [8].

However, the question of how to evaluate the correctness of the exercise is still not certain, as the literature is only represented by a limited number of articles [7,21]. The previous research has demonstrated a most accurate posture classification of 91.9%, and for movement, a most accurate posture classification of 95.16% [21]. This study demonstrated a slight increase in the accuracy by using three different algorithms and by setting up a threshold level for: total error of vector lengths; total error of angles; and multiplication of vector errors by angle errors (as in [21]). Calculating sensitivity and specificity, the classification accuracy of the algorithms was obtained, with the best result shown by the algorithm using the total error of angles (94.9%). This algorithm showed better results when compared with previous research based on a multiplication of the total errors algorithm. This new algorithm also requires considerably fewer parameters for the classification of postures and exercise movements. The previous study, which showed the best accuracy for the posture classification, used 30 variables of the posture descriptor, such as angles and vector lengths [21]. However, the second algorithm in this research used only 17 variables of posture descriptor, which significantly improved the efficiency of the method.

In our study, when evaluating the classification accuracy of the exercises, we used results for the average accuracy of each participant and the average accuracy of the exercises, which were 94.3% (SD 1.7%) and 94.2% (SD 1.8%), respectively. Those results are practically the same as those of the previous research [21], but our algorithm, as mentioned above, requires considerably fewer parameters for the classification of postures and exercise movements. More advanced marker-based motion capture systems can also be used to improve the classification accuracy of algorithms. Previous research [32] has demonstrated that the static error of tracking passive markers with Oqus (Qualisys) cameras was

0.15 mm and a dynamic 0.26 mm, with much higher tracking frequencies than those used by the Kinect V2 sensor.

The definition of human posture can be applied not only to the creation of applications for rehabilitation, but also for monitoring the lives of older people, such as in the recording of a sudden fall. According to statistics, 28–35% of people over 65 years of age experience a fall [33], after which they often need a period of rehabilitation. Such a monitoring system could detect a person's posture, and alert relatives, neighbors or close friends in cases where the person's positional data indicates the possibility of a heart attack, stroke or other complication; such a posture, for example, could be lying down on the floor. The time factor in attending to such situations is very crucial, being directly correlated to the person's recovery.

More studies are required to develop classification algorithms for the various medical applications mentioned, as this study had a number of limitations, outlined below.

1. Limited tested sample size and reference database for healthy subjects.
2. Healthy and young subjects were recruited without any disabilities.
3. Different races, nationalities and type of disability may influence the results, as well as affect anthropometric data.
4. Kinect sensors are not consistent in data collection for different environments, and different types of clothing can significantly change the accuracy of the detection of joints, as was noticed in our study.

Future planned research is to use the Qualisys system to improve the algorithm by reducing the number of limitations.

Video analysis is widely applied in the context of human movement detection, and real-time implementation using reliable algorithms based on the postural recognition of healthy persons should provide postural data that can be used to assess the effectiveness of clinically prescribed exercise regimes for patients, as well as allow for variations in exercise regime, dependent on the data collected. Such data would be useful in optimized treatment by exercise therapy.

The advantages of such an approach could also be extended to veterinary applications. Very few studies address automatic video-based analysis of animals—for example, canine behavior as a means of monitoring animal health and wellbeing [28–30]—with some of these studies using a 3D Kinect camera to detect joint position. In [28], the authors present a system capable of identifying static postures for canines that does not rely on hand-labeled data at any point, although the system can only identify the "standing," "sitting" and "lying" postures with approximately 70%, 69% and 94% accuracy, respectively. Paper [29] presents a depth-based tracking system for the automatic detection of animals' postures and body segments, as well as an exhaustive evaluation on the performance of several classification algorithms, based on both a supervised and a knowledge-based approach. Furthermore, Barnard et al. addressed a problem of automatic behavioral analysis of kenneled dogs using 3D video monitoring [30]. Dog body segment detection was done using standard Structural Support Vector Machine classifiers, and the automatic tracking of the dog was also implemented. However, this tool has a high margin for improvement.

A number of studies were also found in the literature using wide-ranging applications in the biomechanics of animals, as well as in prosthetics to prevent injuries, monitoring rehabilitation after surgical operations, choosing the appropriate orthopedic devices and prostheses, training and others [34–36]. Therefore, the classification algorithm of posture can also be useful in not only human medicine, but also veterinary applications, influencing veterinary intervention using exercise regimes, as well as monitoring animals' health and behavior. Further studies using the Qualisys system and neural network, which would be trained to recognize a dog's skeleton using cost-effective video cameras, are planned; so far, such work has only been carried out for humans.

5. Conclusions

Virtual or home rehabilitation using modern technologies can improve health and quality of life for many people and animals. The algorithms for posture and movement classification used in this study demonstrated good results using an optical sensor. These algorithms can also be used in other motion capture systems as a simpler and less resource-intensive alternative to machine learning and neural network algorithms, thus increasing accuracy.

The posture and movement classification algorithm may also be used to monitor incidental falls in the elderly population that can be associated with heart failure or a stroke, and initiate a call for help.

As for animals, this technique may also be applied for measuring the time budget of animals, indicating the amount or proportion of time that animals spend in different behaviors as a measure for common ethological and welfare parameters [37].

Author Contributions: Conceptualisation, A.A. and T.K.; methodology, A.S.; software, T.K. and A.S.; validation, A.A., T.K. and D.K.; formal analysis, A.S.; investigation, A.A.; resources, A.S.; data curation, A.S.; writing—original draft preparation, A.A. and D.K.; writing—review and editing, D.K. and A.A.; visualization, A.A. and T.K.; supervision, A.Z. and O.A.M.; project administration, D.K. and A.Z.; and funding acquisition, O.A.M., D.K. and A.Z. All authors have read and agreed to the published version of the manuscript.

Funding: This research was supported by a grant from the Ministry of Science & Technology of Israel and by RFBR according to the research project N 19-57-06007.

Acknowledgments: The authors would like to express their sincere gratitude to the Ilizarov Scientific Center for Restorative Traumatology and Orthopaedics for supporting the project.

Conflicts of Interest: The authors declare no conflict of interest.

References

1. Eleni, K.; Srinivas, A.; Judith, J. The aging population: Demographics and the biology of aging. *Periodontol. 2000* **2016**, *72*, 13–18. [CrossRef]
2. Goulding, M.R.; Rodgers, M.E. Trends in aging—United states and worldwide. *MMWR Morb. Mortal. Wkly. Rep.* **2003**, *52*, 101–104.
3. Kirk-Sanchez, N.J.; McGough, E.L. Physical exercise and cognitive performance in the elderly: Current perspectives. *Clin. Interv. Aging* **2014**, *9*, 51–62. [CrossRef] [PubMed]
4. Zhou, Z.; Hou, Y.; Lin, J.; Wang, K.; Liu, Q. Patients' views toward knee osteoarthritis exercise therapy and factors influencing adherence—A survey in china. *Physician Sportsmed.* **2018**, *46*, 221–227. [CrossRef] [PubMed]
5. Lawford, B.J.; Delany, C.; Bennell, K.L.; Hinman, R.S. "I was really sceptical...But it worked really well": A qualitative study of patient perceptions of telephone-delivered exercise therapy by physiotherapists for people with knee osteoarthritis. *Osteoarthr. Cartil.* **2018**, *26*, 741–750. [CrossRef] [PubMed]
6. Tao, G.; Archambault, P.S.; Levin, M.F. Evaluation of kinect skeletal tracking in a virtual reality rehabilitation system for upper limb hemiparesis. In Proceedings of the 2013 International Conference on Virtual Rehabilitation (ICVR), Philadelphia, PA, USA, 26–29 August 2013; pp. 164–165. [CrossRef]
7. Su, C.-J.; Chiang, C.-Y.; Huang, J.-Y. Kinect-enabled home-based rehabilitation system using dynamic time warping and fuzzy logic. *Appl. Soft Comput.* **2014**, *22*, 652–666. [CrossRef]
8. Fern'ndez-Baena, A.; Susín, A.; Lligadas, X. Biomechanical validation of upper-body and lower-body joint movements of kinect motion capture data for rehabilitation treatments. In Proceedings of the 2012 Fourth International Conference on Intelligent Networking and Collaborative Systems, Bucharest, Romania, 19–21 September 2012; pp. 656–661. [CrossRef]
9. Lin, T.; Hsieh, C.; Lee, J. A kinect-based system for physical rehabilitation: Utilizing tai chi exercises to improve movement disorders in patients with balance ability. In Proceedings of the 2013 7th Asia Modelling Symposium, Hong Kong, China, 23–25 July 2013; pp. 149–153. [CrossRef]
10. Lange, B.; Koenig, S.; McConnell, E.; Chang, C.; Juang, R.; Suma, E.; Bolas, M.; Rizzo, A. Interactive game-based rehabilitation using the microsoft Kinect. In Proceedings of the 2012 IEEE Virtual Reality Workshops (VRW), Costa Mesa, CA, USA, 4–8 March 2012; pp. 171–172. [CrossRef]

11. Antón, D.; Goñi, A.; Illarramendi, A.; Torres-Unda, J.J.; Seco, J. Kires: A kinect-based telerehabilitation system. In Proceedings of the 2013 IEEE 15th International Conference on e-Health Networking, Applications and Services (Healthcom 2013), Lisbon, Portugal, 9–12 October 2013; pp. 444–448. [CrossRef]
12. Clark, R.A.; Vernon, S.; Mentiplay, B.F.; Miller, K.J.; McGinley, J.L.; Pua, Y.H.; Paterson, K.; Bower, K.J. Instrumenting gait assessment using the kinect in people living with stroke: Reliability and association with balance tests. *J. Neuroeng. Rehabil.* **2015**, *12*, 15. [CrossRef]
13. Webster, D.; Celik, O. Systematic review of kinect applications in elderly care and stroke rehabilitation. *J. Neuroeng. Rehabil.* **2014**, *11*, 108. [CrossRef]
14. Pastor, I.; Hayes, H.A.; Bamberg, S.J.M. A feasibility study of an upper limb rehabilitation system using kinect and computer games. In Proceedings of the 2012 Annual International Conference of the IEEE Engineering in Medicine and Biology Society, San Diego, CA, USA, 28 August–1 September 2012; pp. 1286–1289. [CrossRef]
15. Saini, S.; Rambli, D.R.A.; Sulaiman, S.; Zakaria, M.N.; Shukri, S.R.M. A low-cost game framework for a home-based stroke rehabilitation system. In Proceedings of the 2012 International Conference on Computer & Information Science (ICCIS), Kuala Lumpur, Malaysia, 12–14 June 2012; pp. 55–60. [CrossRef]
16. Shin, J.-H.; Ryu, H.; Jang, S.H. A task-specific interactive game-based virtual reality rehabilitation system for patients with stroke: A usability test and two clinical experiments. *J. Neuroeng. Rehabil.* **2014**, *11*, 32. [CrossRef]
17. González-Ortega, D.; Díaz-Pernas, F.J.; Martínez-Zarzuela, M.; Antón-Rodríguez, M. A kinect-based system for cognitive rehabilitation exercises monitoring. *Comput. Methods Programs Biomed.* **2014**, *113*, 620–631. [CrossRef] [PubMed]
18. Chang, Y.J.; Han, W.Y.; Tsai, Y.C. A kinect-based upper limb rehabilitation system to assist people with cerebral palsy. *Res. Dev. Disabil.* **2013**, *34*, 3654–3659. [CrossRef] [PubMed]
19. Lozano-Quilis, J.-A.; Gil-Gómez, H.; Gil-Gómez, J.-A.; Albiol-Pérez, S.; Palacios-Navarro, G.; Fardoun, H.M.; Mashat, A.S. Virtual rehabilitation for multiple sclerosis using a kinect-based system: Randomized controlled trial. *JMIR Serious Games* **2014**, *2*, e12. [CrossRef] [PubMed]
20. Venugopalan, J.; Cheng, C.; Stokes, T.H.; Wang, M.D. Kinect-based rehabilitation system for patients with traumatic brain injury. In Proceedings of the 2013 35th Annual International Conference of the IEEE Engineering in Medicine and Biology Society (EMBC), Osaka, Japan, 3–7 July 2013; pp. 4625–4628. [CrossRef]
21. Anton, D.; Goni, A.; Illarramendi, A. Exercise recognition for kinect-based telerehabilitation. *Methods Inf. Med.* **2015**, *54*, 145–155. [CrossRef] [PubMed]
22. Roh, J.; Park, H.-J.; Lee, K.J.; Hyeong, J.; Kim, S.; Lee, B. Sitting Posture Monitoring System Based on a Low-Cost Load Cell Using Machine Learning. *Sensors* **2018**, *18*, 208. [CrossRef] [PubMed]
23. Kim, Y.M.; Son, Y.; Kim, W.; Jin, B.; Yun, M.H. Classification of Children's Sitting Postures Using Machine Learning Algorithms. *Appl. Sci.* **2018**, *8*, 1280. [CrossRef]
24. Giacomozzi, C.; D'Ambrogi, E.; Uccioli, L.; Macellari, V. Does the thickening of achilles tendon and plantar fascia contribute to the alteration of diabetic foot loading? *Clin. Biomech.* **2005**, *20*, 532–539. [CrossRef]
25. Baratloo, A.; Hosseini, M.; Negida, A.; El Ashal, G. Part 1: Simple Definition and Calculation of Accuracy, Sensitivity and Specificity. *Emergency (Tehran, Iran)* **2015**, *3*, 48–49.
26. Mousavi Hondori, H.; Khademi, M. A review on technical and clinical impact of microsoft kinect on physical therapy and rehabilitation. *J. Med. Eng.* **2014**, *2014*, 846514. [CrossRef]
27. Burdea, G.C. Virtual rehabilitation—Benefits and challenges. *Methods Inf. Med.* **2003**, *42*, 519–523.
28. Mealin, S.; Dom'ınguez, I.X.; Roberts, D.L. Semi-supervised classification of static canine postures using the microsoft Kinect. In Proceedings of the Third International Conference on Animal-Computer Interaction, Milton Keynes, UK, 15–17 November 2016; ACM: New York, NY, USA, 2016; p. 16.
29. Pons, P.; Jaen, J.; Catala, A. Assessing machine learning classifiers for the detection of animals behavior using depth-based tracking. *Expert Syst. Appl.* **2017**, *86*, 235–246. [CrossRef]
30. Barnard, S.; Calderara, S.; Pistocchi, S.; Cucchiara, R.; Podaliri-Vulpiani, M.; Messori, S.; Ferri, N. Quick, accurate, smart: 3d computer vision technology helps assessing confined animals behaviour. *PLoS ONE* **2016**, *11*, e0158744. [CrossRef] [PubMed]
31. Psota, E.T.; Mittek, M.; Pérez, L.C.; Schmidt, T.; Mote, B. Multi-Pig Part Detection and Association with a Fully-Convolutional Network. *Sensors* **2019**, *19*, 852. [CrossRef] [PubMed]

32. Feng, Y.; Max, L. Accuracy and precision of a custom camera-based system for 2d and 3d motion tracking during speech and nonspeech motor tasks. *J. Speech Lang. Hear. Res. JSLHR* **2014**, *57*, 426–438. [CrossRef] [PubMed]
33. Vieira, E.R.; Palmer, R.C.; Chaves, P.H.M. Prevention of falls in older people living in the community. *BMJ* **2016**, *353*, i1419. [CrossRef] [PubMed]
34. Prankel, S.; Corbett, M.; Bevins, J.; Davies, J. Biomechanical analysis in veterinary practice. *Practice* **2016**, *38*, 176–187. [CrossRef]
35. Farrell, B.J.; Prilutsky, B.I.; Kistenberg, R.S.; Dalton, J.F.T.; Pitkin, M. An animal model to evaluate skin-implant-bone integration and gait with a prosthesis directly attached to the residual limb. *Clin. Biomech.* **2014**, *29*, 336–349. [CrossRef]
36. Druen, S.; Boddeker, J.; Meyer-Lindenberg, A.; Fehr, M.; Nolte, I.; Wefstaedt, P. Computer-based gait analysis of dogs: Evaluation of kinetic and kinematic parameters after cemented and cementless total hip replacement. *Vet. Comp. Orthop. Traumatol. VCOT* **2012**, *25*, 375–384. [CrossRef]
37. Arney, D. What is animal welfare and how is it assessed? In *Sustainable Agriculture*; Baltic University Press: Uppsala, Sweden, 2012; pp. 311–315.

© 2020 by the authors. Licensee MDPI, Basel, Switzerland. This article is an open access article distributed under the terms and conditions of the Creative Commons Attribution (CC BY) license (http://creativecommons.org/licenses/by/4.0/).

Article

Three-Dimensional (3D) Model-Based Lower Limb Stump Automatic Orientation

Dmitry Kaplun [1,2], Mikhail Golovin [3], Alisa Sufelfa [1,3,*], Oskar Sachenkov [4], Konstantin Shcherbina [3], Vladimir Yankovskiy [3], Eugeniy Skrebenkov [3], Oleg A. Markelov [2] and Mikhail I. Bogachev [2]

1. Department of Automation and Control Processes, Saint Petersburg Electrotechnical University "LETI", 5 Professor Popov Street, 197376 Saint Petersburg, Russia; dikaplun@etu.ru
2. Centre for Digital Telecommunication Technologies, Saint Petersburg Electrotechnical University "LETI", 5 Professor Popov Street, 197376 Saint Petersburg, Russia; oamarkelov@etu.ru (O.A.M.); mibogachev@etu.ru (M.I.B.)
3. Albrecht Federal Scientific Center of Rehabilitation of the Disabled, 50 Bestuzhevskaya Street, 195067 Saint Petersburg, Russia; golovin@center-albreht.ru (M.G.); reabin@center-albreht.ru (K.S.); yankovsky.vladimir@yandex.ru (V.Y.); skrebenkovea@center-albreht.ru (E.S.)
4. Institute of Mathematics and Mechanics, Kazan Federal University, 18 Kremlyovskaya Street, 420008 Kazan, Russia; 4works@bk.ru
* Correspondence: sufelfaar@center-albreht.ru; Tel.: +7812-544-34-66

Received: 28 March 2020; Accepted: 30 April 2020; Published: 7 May 2020

Abstract: Modern prosthetics largely relies upon visual data processing and implementation technologies such as 3D scanning, mathematical modeling, computer-aided design (CAD) tools, and 3D-printing during all stages from design to fabrication. Despite the intensive advancement of these technologies, once the prosthetic socket model is obtained by 3D scanning, its appropriate orientation and positioning remain largely the responsibility of an expert requiring substantial manual effort. In this paper, an automated orientation algorithm based on the adjustment of the 3D-model virtual anatomical axis of the tibia along with the vertical axis of the rectangular coordinates in three-dimensional space is proposed. The suggested algorithm is implemented, tested for performance and experimentally validated by explicit comparisons against an expert assessment.

Keywords: 3D model; prosthetic design; orientation; positioning; reconstruction

---

1. Introduction

According to World Health Organization (WHO) statistics, up to one billion people in the world constituting about 15% of the total population live with certain form of disability, including approximately 200 million experiencing considerable difficulties in functioning that limit their participation in family and society life, with the domination of cases in the upper age group of ≥60 years old, where nearly every second a person experiences moderate and about 10% severe disabilities [1]. Severe disabilities are often associated with lower limb amputations, with approximately 30,000 to 40,000 amputations performed each year, and more than 1.5 million people living with a lost limb in the U.S. alone, with the majority requiring access to lower limb prosthetics [2]. The total number of lower limb prostheses provided annually in Russia increased from around 60,000 in 2012–2014 to above 80,000 in 2016–2018 [3].

Although reasons for amputation vary considerably between regions and age groups, ranging from severe trauma, wounds and burns caused by road traffic and occupational injuries as well as violence and humanitarian crises to tumors and inflammatory diseases potentially leading to the development of sepsis and multiple organ failure, on average around one half of amputation cases can be attributed to either trauma or disease, respectively [1,4].

Modern prosthetics largely rely upon technological support to improve the accuracy of design and finally to enhance the quality of rehabilitation taking into account the individual characteristics of the patient. Current technologies allow for a highly automated prosthetic design procedures based on the geometry and properties of human organs obtained by 3D scanning [5] or computed tomography (CT) scanning [6] followed by prosthetic model reconstruction. In addition to the design itself, adequate 3D models allow for running biomechanical simulations, leading to certain optimization strategies based on objective visual data sources and numerical mathematics [7]. In some cases, appropriate model design requires a combination of different technologies. In contrast to 3D scanning, CT and magnetic resonance imaging (MRI) are also capable of visualization of soft tissues [8,9]. Thus, modern scanners with the usage of digital models and mathematical simulations can predict the behavior of patient's organs at both micro [10] and macro scales [11] as well as over different time spans [12].

Modern approach to the design of receiving sockets for lower limb prostheses typically consists of the following stages:

1. 3D scanning of the lower limb stump;
2. Design and analysis of a digital geometric model;
3. Printing the prosthetic socket using a 3D printer.

The 3D scanning involves recording of the surface points coordinates [13] using a 3D scanner or recording device followed by the reconstruction procedure used to obtain a 3D-model of a limb segment. One of the key stages in 3D-model processing is the orientation of the model in 3D Cartesian coordinates.

Among recent approaches to the automation of the 3D-model orientation problem, using various mathematical filters for reconstructions of the 3D-model from 2D X-ray images can be mentioned [14]. The rotation matrix used for the automatic orientation of the model is in 3D space. However, this approach is limited to certain specific 3D-model rotation angles.

Another approach is based on the virtual environment that is used to automatically select the position of the 3D-model in three-dimensional space is considered in [15]. The key disadvantage of this particular method is that it limits possible 3D-model orientations to a set of specified templates.

In another recent work [16] orientation of the 3D image of the scapula is based on the preliminarily fixed markers that are being further compared against the control model. The disadvantage of this method is that the expert needs to place markers in the 3D-model manually.

A commonly applied solution is a web-based program, Rapid Plaster (PVA Med, New York, NY, USA), developed for the design of prosthetic sockets allowing users to work with a 3D-model in a conventional browser and convert it using standard tools for working with 3D models, although again it lacks a component that could be used for the automated orientation in 3D space [17].

Current classification of socket types, their common advantages and limitations from multiple viewpoints, as well as an overview of the keynote parameters affecting the stump–socket interface and influencing the comfort and stability of the limb prostheses such as displacement, stress, volume and temperature fluctuations for different positioning and orientation scenarios are reviewed in [18].

A physically motivated reduced model of the dynamic interactions between the residual limb and the prosthetic socket proposed in [19] is capable of the quantitative assessment and simulation of the stress distribution for different variants of the socket orientation and positioning as well as its possible alterations depending on the corresponding variation in the friction coefficients.

Another recent knowledge-based approach to the design and orientation of the lower limb prostheses with particular focus on the 3D modeling of the socket is based on the acquisition and formalization of the knowledge related to the prosthesis manufacturing process as well as the architecture of a dedicated knowledge-based engineering framework detailing the key design steps. A computer-aided module named socket modeling assistant (SMA) represents a virtual laboratory where the socket prototype is being created and positioned based on the digital model of a patient's residual limb. The SMA software acts as an interactive tool to guide and support the expert socket designer during each step from socket design to its positioning and orientation either in an automatic or in a semi-automatic fashion [20].

The authors of [21] focus on untangling the complexity of the transtibial prosthetic socket fit by selecting the keynote characteristics that guarantee its successful fitting as well as finding certain criteria for the optimized selection of the particular prosthetic socket type for different positioning and orientation scenarios. Based on the analysis of the activity levels and reported satisfaction of active persons, especially among younger persons mainly with a traumatic cause of amputation, they conclude that the total surface-bearing sockets generally outperform the patellar tendon-bearing sockets, as they can be better adjusted to the lower limb stump in order to withstand dynamic loads typically associated physical activity.

Socket biomechanics, socket pressure measurement, friction-related phenomena and associated properties, with corresponding computational models describing the limb tissue responses to the external mechanical loads and other physical conditions for different socket positioning and orientation scenarios are the key focus of [22]. Further advancement of this research is associated with the embedded sensors enabling direct measurements of the physical stresses applied to the socket [23]. The results of this study indicate that the direction and the angle of rotation of the stump could be obtained by decoding the magnetic field signals obtained by magnetic sensors embedded into a prosthetic socket. This pilot study provides important guidelines for the development of a practical interface between the residual bone rotation and the prosthesis for control of prosthetic rotation.

The conventional approach to the prosthetic socket orientation includes splitting the stump 3D-model into sections according to a discrete grid in the horizontal plane followed by obtaining the resulting dimensions from the cross-section perimeters. Accordingly, registration of the cross-section perimeters while the 3D model is positioned at the wrong angle to the horizontal plane leads to incorrect measurements. Thus, it is essential to create an algorithm that provides at least preliminary automated orientation of the 3D-model based on objective measurement data. Due to high variability in the stump characteristics and the individual shape of each stump, especially in each congenital malformation case, in certain scenarios only preliminary conclusions can be made automatically, and expert attention will, nevertheless, be required. However, even in this scenario, an automatic decision support system (DSS) could save an expert's time and provide at least preliminary positioning.

In this work, we propose an automated orientation algorithm based on the adjustment of the 3D-model virtual anatomical axis of the tibia along with the vertical axis of the rectangular coordinates in three-dimensional space. The suggested algorithm is implemented, tested for performance and experimentally validated by explicit comparisons against an expert assessment. We believe that the proposed solution could be useful as a DSS for the prosthetic expert support.

The automated system development is based on the algorithm of orientation of the digital geometric model of the lower limb stump in an automatic mode, and consists of the following steps:

1. Design of the 3D-model of the lower limb stump obtained by 3D scanning of the patient.
2. Development of a decision support system for the automatic 3D-model orientation.
3. Comparative analysis of the model-based predictions with expert evaluation results.
4. Estimation of the algorithm performance including the duration of automatic orientation for a given resolution setting associated with the number of 3D-model polygons.

2. Materials and Methods

The anonymous experimental data for the study were obtained from the Federal Scientific Center of Rehabilitation of the Disabled named after G.A. Albrecht. The study was performed in accordance with the ethical standards presented in the Declaration of Helsinki. The study protocol was reviewed and approved by the local expert collegiate council before the beginning of the study. All patients provided their written informed consent prior to their inclusion in the study.

The study focused on the patients with unilateral, bilateral and multiple amputation defects with various causes of amputation defects, with one lower limb having one defect at the lower leg level. The following patients were excluded from enrollment in the study: those with skin defects such as non-healing wounds; those with trophic ulcers; those exhibiting tremor of the extremities due to various diseases; those with protrusion of bone sawdust under the skin that prevents prosthetics; those suffering from either acute myocardial infarction or acute cerebrovascular accident within 4 months prior to the study; those exhibiting mental illness in the acute stage; those with contagious infectious diseases; and those with complication of the main and/or concomitant disease with the appointment of bed rest.

Among seven patients selected for detailed investigation, four were male and three female, aged between 45 and 58 years old.

Mathematical analysis and functional programming tools (Python 3.3.7 with Mesh library, www.python.org) were used for visual data processing, model design and computer simulations. The loop optimization method was used to select the points with maximum displacement [24]. Multidimensional regression method was used for 3D-model cross-section fitting [25]. For 3D-model orientation, quaternions [26] and rotation matrices have been used [27].

The developed algorithm (see Algorithm 1) performs an automatic orientation of the 3D-model. The key steps of the algorithm are as follows:

Algorithm 1

---

1. Data import (e.g., reading from file)
2. Selection of points with maximum coordinates (loop optimization method)
3. Primary 3D-model spatial orientation
4. Setting cross-section under control by the operator
5. Arrangement of the points defining the cross-section
6. Finding cross-section central points
7. Calculation of the virtual anatomical axis
8. Secondary 3D-model spatial orientation
9. Data export (e.g., saving to file)

In the first step a digital representation obtained by a 3D scanner is imported as an array of coordinates containing the points $r_i = (x_i, y_i, z_i)$. Once imported, all entries are sorted according to their coordinates along the Z axis. Figure 1 illustrates the visual representation of the input model, where x, y, z are Cartesian axes, while a is the virtual anatomical axis. The spatial orientation of the model is determined by the angles between the projections of the anatomical axis a on the xy, xz and zy planes.

Next, the points $r_i = (x_i, y_i, z_i)$ characterized by the maximum coordinates (step 2 of the algorithm) have been selected as:

$$\vec{r}_{max} = \max_z \vec{r}_i \qquad (1)$$

$$\vec{r}_{min} = \min_z \vec{r}_i \qquad (2)$$

To enable the rotation of the digital 3D model, we employed a set of transformations following the rotation matrices approach. In 3D space rotation around the Z-axis is described by the matrix transformation.

$$R_z(\theta) = \begin{pmatrix} \cos(\theta) & \sin(\theta) & 0 \\ -\sin(\theta) & \cos(\theta) & 0 \\ 0 & 0 & 1 \end{pmatrix} \qquad (3)$$

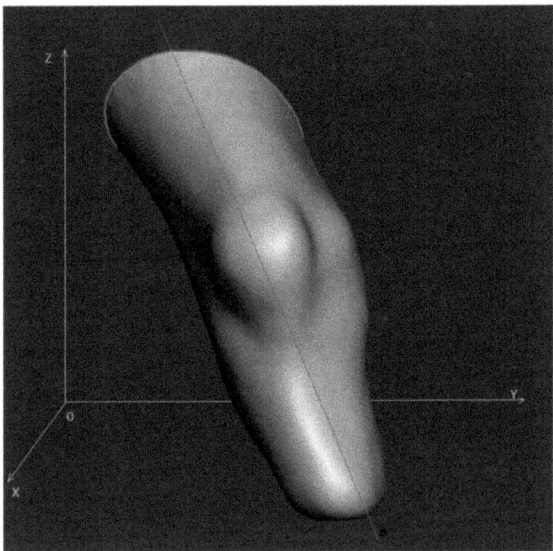

Figure 1. Visual representation of the input model.

Similarly, rotation around the X-axis has the form:

$$R_x(\varphi) = \begin{pmatrix} 1 & 0 & 0 \\ 0 & \cos(\varphi) & \sin(\varphi) \\ 0 & -\sin(\varphi) & \cos(\varphi) \end{pmatrix} \quad (4)$$

Rotation around the Y-axis is given by:

$$R_y(\omega) = \begin{pmatrix} \cos(\omega) & 0 & -\sin(\omega) \\ 0 & 1 & 0 \\ \sin(\omega) & 0 & \cos(\omega) \end{pmatrix} \quad (5)$$

Next, the coordinates of the point r_i can be written as a series of consecutive transformations.

$$r_i^* = R_z(\theta) \cdot R_x(\varphi) \cdot R_y(\omega) \cdot r_i^T \quad (6)$$

The resulting transformation matrix can be interpreted as a matrix of the direction cosines between the old and new coordinate systems.

$$r_i^* = R_z(\theta) \cdot R_x(\varphi) \cdot R_y(\omega) \cdot r_i^T = M \cdot r_i^T \quad (7)$$

The described approach has been used both for the primary and for the secondary orientation of the model (steps 3 and 8 of the algorithm).

The normalized vector indicating the direction that connects these points relative to the global coordinate system r_{dir} can be expressed as:

$$\vec{r}_{dir} = \frac{\vec{r}_{max} - \vec{r}_{min}}{\|\vec{r}_{max} - \vec{r}_{min}\|} \quad (8)$$

Figure 2 shows the cross-sections of the 3D-model before (2A) and after (2B) arrangement.

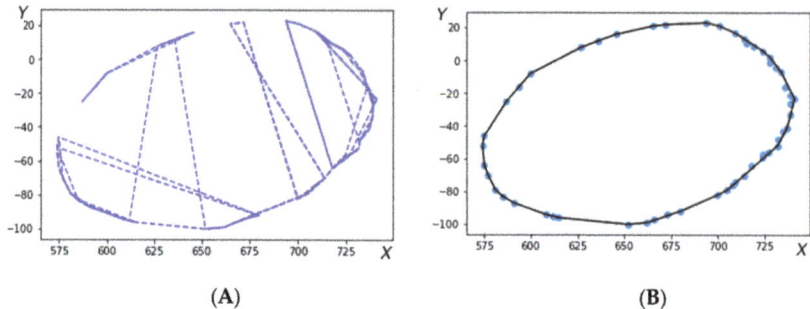

Figure 2. Cross-sections: (**A**) points before arrangement, (**B**) points after arrangement.

Following the arrangement of the points constituting the cross-section a closed contour is selected (step 5 of the algorithm). Creation of a closed contour starts from a reference point and proceedings following the oriented direction obtained at each j-th step from:

$$D_j = \frac{\| r_0 - r_i \| \cdot \| r_0 \| \cdot \| r_i \|}{(r_0, r_i)} \qquad (9)$$

Arrangement of this array according to a specific variable D_j, defines an ordered closed set of points, as exemplified in Figure 2b, where r is an array of coordinates containing the points $r_i = (x_i, y_i, z_i)$.

At step 6, the central (pivot) points are obtained by averaging the coordinate points for each of the cross sections specified by the operator. Next at step 7 a straight line through the central points (obtained at step 6) is fitted using the multidimensional regression method. In particular, the error ε is being minimized in the model:

$$Y = Zb + d + \varepsilon \qquad (10)$$

where Y contains z-coordinates from the dataset r, Z contains x- and y-coordinates from the dataset r, b and d are free model parameters, while ε is the random error. The problem (10) can be resolved using the least mean squares (LMS) method.

$$S = \sum (Y - Zb - d)^2 \to \min \qquad (11)$$

thus yielding.

$$\begin{cases} \sum (Y - Zb - d)Z = 0 \\ \sum (Y - Zb - d) = 0 \end{cases} \qquad (12)$$

Finally, the cross-section levels of the lower limb stump in the 3D-model are obtained, and the entire 3D-model is aligned along the Z-axis (see Figure 3) before being exported.

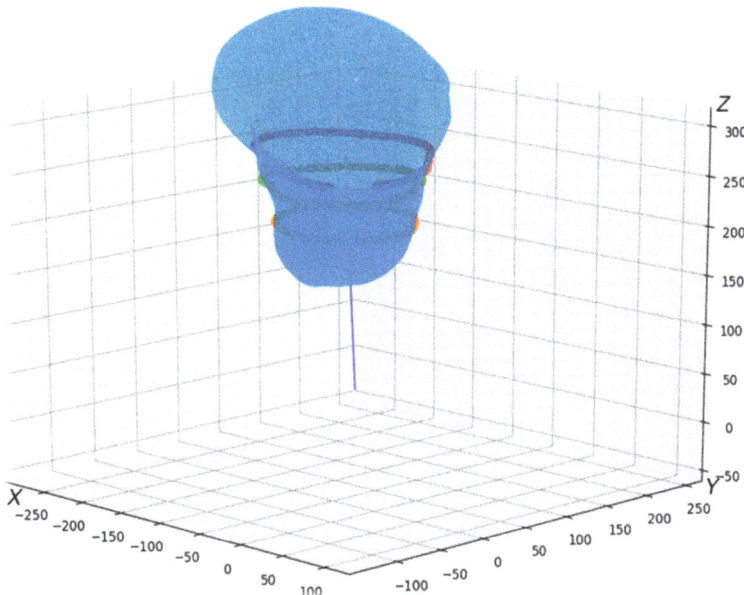

Figure 3. Three-dimensional (3D) model with cross-sections and fitting line that passed through its centers.

Figure 4 shows the model appearance after the orientation using the designed algorithm.

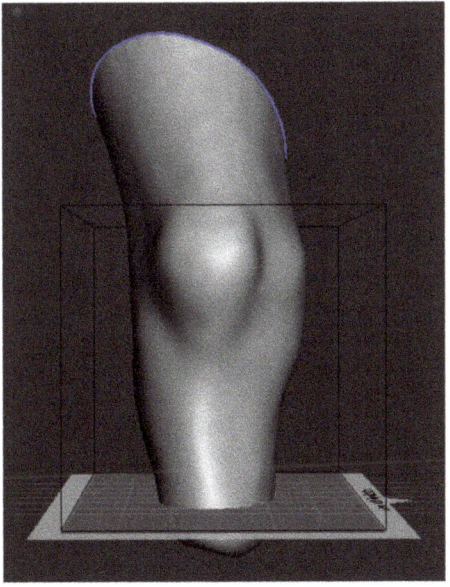

Figure 4. Visual representation of the oriented 3D model.

3. Results

The validation of the proposed algorithm was performed using experimental data obtained by 3D scanning followed by 3D model reconstructions for seven different patient cases as summarized in Table 1. To validate the tests under different initial conditions, prior to the test each 3D model was randomly oriented. Since there is no "gold standard" for an automated algorithm in the field, the results of the automatic orientation algorithm were compared against manual positioning by an experienced prosthetic expert.

Table 1. Summary of validation results based on experimental measurements from seven different patients compared against prosthetics expert assessment.

Case No	Cross Sections	Measurement by an Automated Algorithm, mm	Measurement by a Prosthetics Expert, mm	Difference between the Results, mm
1	1 cross section	149.800	157.847	8.046
	2 cross section	302.437	317.164	14.727
	3 cross section	206.625	206.369	0.256
2	1 cross section	No reasonable result has been obtained	42.663	-
	2 cross section		322.976	-
	3 cross section		119.738	-
3	1 cross section	265.208	254.661	10.547
	2 cross section	371.080	376.716	5.636
	3 cross section	476.239	477.109	0.870
4	1 cross section	154.229	143.349	10.880
	2 cross section	261.332	260.572	0.760
	3 cross section	408.356	412.207	3.850
5	1 cross section	65.501	63.332	2.169
	2 cross section	359.217	359.341	0.125
	3 cross section	244.260	224.179	20.081
6	1 cross section	44.841	54.874	10.032
	2 cross section	215.699	222.156	6.458
	3 cross section	345.907	342.486	3.421
7	1 cross section	101.854	103.752	1.898
	2 cross section	366.239	364.057	2.182
	3 cross section	460.786	461.043	0.257

For the quantitative comparison between the automated and the expert manual positioning, two pivot points, one on the distal and one on the proximal planes, were selected, and a straight line connecting these two points was fitted. Once the model orientation was performed independently by the automated algorithm and manually by the expert, the two spatially oriented models were matched by the point in the distal plane. Three projections of the straight line on the xy, xz and zy planes were obtained for both automated and manually oriented models, and angles between these projections in each plane were used for the quantitative comparison between the two approaches, as summarized in Table 1. In the above procedure, the exact locations of the pivot points are not critical for the accuracy of the comparison, since they are used for measuring relative orientations only.

Although in this work we focused on the lower limb stump orientation problem, the proposed algorithm appears to be a universal tool for the 3D model orientation. We have implemented the algorithm using Python programming language and Mesh library features (the source code is available as Supplementary Material). The developed software module follows the algorithm represented below in pseudocode (see Algorithm 2). The algorithm contains two key verification conditions, with the first condition checking whether the virtual anatomical axis of the tibia coincides with the vertical Z-axis, while the second condition triggering if the model is positioned with the stump end up after orientation.

Since the construction of the cross-sections depends explicitly on the anatomic characteristics of the lower limb stump, it is specified by the operator for each 3D-model in an individual range.

The axis of the model should coincide with the Z-axis, since this orientation is required for a more intuitive manual positioning by the expert, who is more used to traditional technology assuming the adjustment of the stump cast installed vertically. After the manual adjustment is performed, the entire model can be re-positioned and re-oriented as required to match with the actual biomechanical axis.

The angle of the virtual axis along the tibia to the vertical Z-axis is affected by the angle of the 3D-model slice after scanning. Figure 5 compares the models obtained after orientation, where (*a*) is an expert oriented 3D-model, while 3D-models (*b*)-(*f*) are different variant of 3D-model oriented by the proposed algorithm. The 3D-model gradually shortens the end above the knee. The data obtained show the following: the shorter the femoral part of the 3D-model, the smaller the deviation of the virtual anatomical axis of the tibia from the vertical Z-axis.

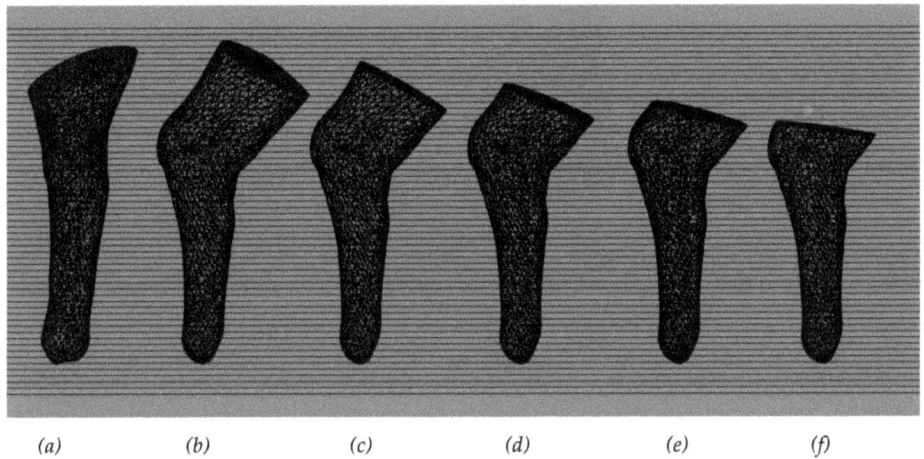

(a)   (b)   (c)   (d)   (e)   (f)

Figure 5. Comparison of models obtained after orientation: (**a**) an expert oriented 3D-model, (**b**–**f**) 3D-models oriented by the proposed algorithm.

Algorithm 2

1. Data import (e.g., reading from file): array Data(3, N), where N–number of points.
2. Selection of points with maximum coordinates: calculate Rmax, Rmin according to Equations (1) and (2)
3. Estimation of the rotation angles: calculate matrix M–array RotMat(3, 3) for direction (Rmin, Rmax) according Equation (8) and recalculate data according to Equation (7)
4. Arrangement points in the cross-section: calculate array DataOrd(3, N) according to Equation (9), calculate central points of sections by averaging CentralData(3,K), where K–number of sections
5. Fitting a regression line: for CentralData calculate vectors bVec and dVec according to Equation (12)
6. Estimation of the rotation angles: calculate RotMat(3, 3) for direction bVec and recalculate Data according to Equation (7)
7. **If** 3D model axis does not coincide with Z axis, **then go to** line 2
8. Setting cross-section under control by the operator
9. 3D model centering
10. **If** 3D model inverted, **then** flip over the model
11. Data export (e.g., saving to file)

Consequently, the input requirements for the algorithm performance are: (i) 3D-model scan length above the knee and (ii) the angle of inclination of the cross-section plane of the scan above the knee to the anatomical axis of the femur.

For comparison the inclination angle input models can be marked with reference points. To analyze the angles of deviation of the virtual anatomical axis of the tibia from the Z ax is after orientation by the algorithm and the expert. In Figure 6, a straight line drawn through reference points allows the results to be compared visually.

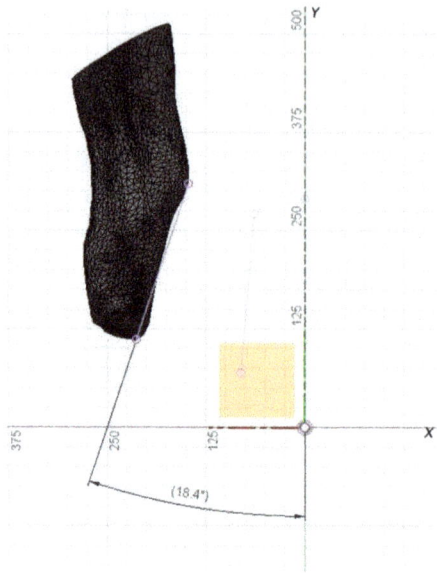

Figure 6. Inclination angle calculation.

Performance of the algorithms depends explicitly on the number of polygons. The dependence of the analysis time on the number of polygons in the model is shown in Figure 7.

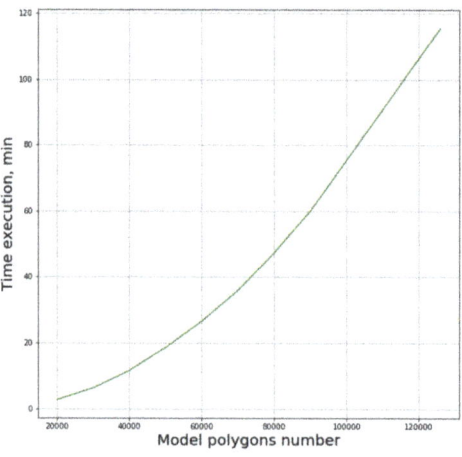

Figure 7. Performance as a function of polygon number in the 3D model.

The presented dependence function can be reasonably fit by a second order polynomial. It was found that significant simplification of the original model leading to the six-fold reduction of the number of polygons from 120,000 to 20,000, the algorithm's operation time is reduced by approximately 1.5 orders of magnitude. Thus, reduction of calculation time can be achieved by any measures that lead to the reduction of the polygon number, such as smoothing of the 3D-model surfaces.

4. Discussion

Based on the results obtained, it can be assumed that the developed algorithm provides with relatively good agreement with the expert assessment typically within 10 degrees along the vertical axis in any direction, as revealed by comparison of the fitting line obtained by the automatic algorithm and a similar line fitted for the model oriented manually by the expert. The most pronounced deviations could be observed in the third section for the fifth patient that could be attributed to the specific anatomic features of the stump. As indicated in Table 1, no relevant results were obtained for the second patient, as attempted analysis resulted in the inverted positioning of the 3D model. A possible reason for this result is that the angle of inclination of the 3D-model cross section plane above the knee to the virtual anatomical axis of the femur was untypically sharp, resulting in the input requirements not being fulfilled.

The limitations of the proposed approach include limited automation, as the expert has to define the position of the proximal edge of the stump to determine the location of the upper section. We also have to note that the algorithm in its present implementation relies upon the assumption that the virtual anatomic axis is orthogonal to the sagittal plane, which is not exactly, but only approximately fulfilled especially for longer models, see also Figure 6 for visual reference.

In our opinion, the above algorithm, taking into account all its inherent limitations, can nevertheless be used as a decision support tool for the prosthetic expert and 3D designer considerably reducing the workload of qualified personnel. Among relevant analogues, the work [28] should be mentioned, where 2D cross-sections of the upper and lower limbs 3D-model were used to create a "bone structure". The sections were created by placement of any number of parallel planes inside the 3D-model and calculating corresponding intersections between the 3D-model and the plane. This "bone structure" can be used to determine the orientation of the scanned model in 3D space in order to create additional cross-section images, as well as obtaining basic information about the shape of the hand or foot or missing fingers, which can be used to create a 3D-model of the prosthesis. However, getting accurate dimensional information about the 3D-model being scanned will require additional input data from the user to determine the scale of the 3D-model, which in our opinion appears to be a certain limitation of this approach, as preprocessing of the 3D-model by an expert is required increasing the workload and potentially reducing the objectivity of the first approach to the model orientation.

As an outlook, we believe that further optimization of the algorithm could make the dependence of the performance on the number of polygons in the 3D-model less pronounced. For that, an algorithm module that allows orientation of the 3D-model in a horizontal plane using the quaternion method for rotating an array of 3D-model points sounds a perspective solution. Since recent technological advancement has enabled the estimation and simulation of the distribution of stresses applied to the residual limb tissues, as well as the volume fluctuations affecting the stump over time and the temperature variations influencing the residual tissues for different variants of socket design, a more automated functional socket design may be developed in the near future. Moreover, recent advancements in prosthetic technologies suggest that future directions could be associated with advanced socket designs capable of self-adaptation to the complex interplay of factors affecting the stump, under both static and dynamic loads, as a replacement for the current fixed socket-orientation scenario [18].

5. Conclusions

To summarize, an automated orientation algorithm based on the adjustment of the 3D-model virtual anatomical axis of the tibia along with the vertical axis of the rectangular coordinates in three-dimensional space was proposed. The suggested algorithm was implemented, tested for performance, and experimentally validated by explicit comparisons against expert assessment. Based on the results obtained, it can be assumed that the developed algorithm provides relatively good agreement with the expert assessment typically within 10 degrees along the vertical axis in any direction, taking into account the input requirements. In our opinion, the above algorithm, taking into account all its inherent limitations, can nevertheless be used as a decision support tool for the prosthetic expert and 3D designer considerably reducing the workload of qualified personnel.

Supplementary Materials: The following are available online at http://www.mdpi.com/2076-3417/10/9/3253/s1: Python Software Module S1: orientation script.

Author Contributions: Conceptualization, M.G. and V.Y.; methodology, O.S.; software, A.S.; validation, E.S., M.G. and K.S.; formal analysis, V.Y., E.S.; investigation, A.S., M.G.; resources, K.S. and O.A.M.; data curation, O.S.; writing—original draft preparation, A.S., D.K.; writing—review and editing, D.K., M.I.B.; visualization, A.S. and O.A.M.; supervision, D.K.; project administration, M.I.B.; funding acquisition, M.I.B. All authors have read and agreed to the published version of the manuscript.

Funding: This research was funded by the Ministry of Science and Higher Education of the Russian Federation under assignment No. 0788-2020-0002. The APC was funded by the Ministry of Science and Higher Education of the Russian Federation under assignment No. 0788-2020-0002.

Acknowledgments: The authors would like to thank the staff of the Federal Scientific Center of Rehabilitation of the Disabled named after G.A. Albrecht and the Regional Scientific and Educational Mathematical Center at Kazan Federal University for assistance.

Conflicts of Interest: The authors declare no conflict of interest.

References

1. World Report on Disability. 2011. Available online: https://www.who.int (accessed on 5 March 2020).
2. American Academy of Physical Medicine and Rehabilitation. Available online: https://www.aapmr.org (accessed on 5 March 2020).
3. Russian Federation Open Data Portal. Available online: https://data.gov.ru/ (accessed on 10 April 2020).
4. Bumbaširević, M.; Lesic, A.; Palibrk, T.; Milovanovic, D.; Zoka, M.; Kravić-Stevović, T.; Raspopovic, S. The current state of bionic limbs from the surgeon's viewpoint. *EFFORT* **2020**, *2*, 5. [CrossRef] [PubMed]
5. Greatrex, F.; Montefiori, E.; Grupp, T.; Kozak, J.; Mazzà, C. Reliability of an Integrated Ultrasound and Stereophotogrammetric System for Lower Limb Anatomical Characterisation. *Appl. Bionics Biomech.* **2017**, *2017*, 4370819. [CrossRef] [PubMed]
6. Sachenkov, O.A.; Gerasimov, O.V.; Koroleva, E.V.; Mukhin, D.A.; Yaikova, V.V.; Akhtyamov, I.F.; Shakirova, F.V.; Korobeynikova, D.A.; Khan, H.C. Building the inhomogeneous finite element model by the data of computed tomography. *Russ. J. Biomech.* **2018**, *22*, 291–303.
7. Gerasimov, O.V.; Berezhnoi, D.V.; Bolshakov, P.V.; Statsenko, E.O.; Sachenkov, O.A. Mechanical model of a heterogeneous continuum based on numerical-digital algorithm processing computer tomography data. *Russ. J. Biomech.* **2019**, *23*, 87–97.
8. Kuchumov, A.G. Biomechanical model of bile flow in the biliary system. *Russ. J. Biomech.* **2019**, *23*, 224–248.
9. Kuchumov, A.G.; Selyaninov, A. Application of computational fluid dynamics in biofluids simulation to solve actual surgery tasks. *Adv. Intell. Syst. Comput.* **2020**, *1018*, 576–580.
10. Marcián, P.; Wolff, J.; Horáčková, L.; Kaiser, J.; Zikmund, T.; Borák, L. Micro finite element analysis of dental implants under different loading conditions. *Comput. Biol. Med.* **2018**, *96*, 157–165. [CrossRef] [PubMed]
11. Ridwan-Pramana, A.; Marcian, P.; Borak, L.; Narra, N.; Forouzanfar, T.; Wolff, J. Finite element analysis of 6 large PMMA skull reconstructions: A multi-criteria evaluation approach. *PLoS ONE* **2017**, *12*, e0179325. [CrossRef]

12. Eggermont, F.; Derikx, L.C.; Verdonschot, N.; Van Der Geest, I.C.M.; De Jong, M.A.A.; Snyers, A.; Van Der Linden, Y.M.; Tanck, E. Can patient-specific finite element models better predict fractures in metastatic bone disease than experienced clinicians. *Bone Jt. Res.* **2018**, *7*, 430–439. [CrossRef]
13. Tinsley Grant, M.; Moore, M.L.; Benavides Marqui, L.; Dellinger Jacob, R.; Adamson Brian, T. 3-Dimensional optical scanning for body composition assessment: A 4-component model comparison of four commercially available scanners. *Clin. Nutr.* **2020**, *3*. [CrossRef]
14. Janssen, M.H.J.; Janssen, A.J.E.; Bekkers, E.J.; Bescós, J.O.; Duits, R.J. Design and processing of invertible orientation scores of 3D images. *Math. Imaging* **2018**, *60*, 1427–1458. [CrossRef] [PubMed]
15. Laga, H. Data-driven approach for automatic orientation of 3D shapes. *Vis. Comput.* **2011**, *27*, 977. [CrossRef]
16. Nicholson, K.F.; Richardson, R.T.; Miller, F.B.; Richards, J.G. Determining 3D scapular orientation with scapula models and biplane 2D images. *Med. Eng. Phys.* **2017**, *41*, 103–108. [CrossRef]
17. Rapidplaster. Available online: https://rapidplaster.pva.net/rapid-plaster (accessed on 15 October 2019).
18. Paterno, L.; Ibrahimi, M.; Gruppioni, E.; Menciassi, A.; Ricotti, L. Sockets for limb prostheses: A review of existing technologies and open challenges. *IEEE Trans. Biomed. Eng.* **2018**, *65*, 1996–2010. [CrossRef]
19. Noll, V.; Eschner, N.; Schumacher, C.; Beckerle, P.; Rinderknecht, S. A physically-motivated model describing the dynamic interactions between residual limb and socket in lower limb prostheses. *Curr. Dir. Biomed. Eng.* **2017**, *3*, 15–18.
20. Rizzi, C.; Colombo, G.; Facoetti, G.; Gabbiadini, S. 3D Modelling of Prosthesis Socket with a Knowledge Based Approach. In Proceedings of the Tools and Methods of Competitive Engineering TMCE 2010, Ancona, Italy, 12–16 April 2010; Volume 1.
21. Safari, M.R.; Meier, M.R. Systematic review of effects of current transtibial prosthetic socket designs—Part 1: Qualitative outcomes. *J. Rehabil. Res. Dev.* **2015**, *52*, 491–508. [CrossRef] [PubMed]
22. Arthur, F.T.; Ming, Z.; David, A.B. State-of-the-art research in lower-limb prosthetic biomechanics socket interface. *J. Rehabil. Res. Dev.* **2001**, *38*, 161–174.
23. Li, G.; Kuiken, T.A. Modeling of prosthetic limb rotation control by sensing rotation of residual arm bone. *IEEE Trans. Biomed.* **2008**, *55*, 2134–2142. [CrossRef] [PubMed]
24. Miller, F.P.; Vandome, A.F.; McBrewster, J. *Loop Optimization Compiler, Control Flow, Compiler Optimization, Loop Interchange, Loop Splitting, Loop Fusion, Loop Fission, Loop Unwinding*; Alphascript Publishing: Saarbrücken, Germany, 2010; pp. 12–37.
25. Kutner, M.; Nachtsheim, C.; Neter, J.; Li, W. *Applied Linear Statistical Models*; McGraw Hill/Irwin: New York, NY, USA, 2004; Volume 5, pp. 49–52.
26. Kuipers, J.B. *Quaternions and Rotation Sequences*; Princeton University Press: Princeton, NJ, USA, 2002; pp. 127–143.
27. Boyd, S. *Introduction to Applied Linear Algebra Vectors, Matrices, and Least Squares*; Cambridge University Press: Cambridge, UK, 2018; Volume 1, pp. 115–191.
28. Kovalovs, M. A method for automatic analysis of scanned 3D models of human hands and feet. In Proceedings of the Optics, Photonics, and Digital Technologies for Imaging Applications, Strasbourg, France, 24–26 April 2018; Volume 5, pp. 1–6.

© 2020 by the authors. Licensee MDPI, Basel, Switzerland. This article is an open access article distributed under the terms and conditions of the Creative Commons Attribution (CC BY) license (http://creativecommons.org/licenses/by/4.0/).

Article

Improving Calculation Accuracy of Digital Filters Based on Finite Field Algebra

Dmitry Kaplun [1], Sergey Aryashev [2], Alexander Veligosha [3], Elena Doynikova [4], Pavel Lyakhov [1,5] and Denis Butusov [1,6,*]

1. Department of Automation and Control Processes, Saint Petersburg Electrotechnical University "LETI", 197376 Saint Petersburg, Russia; dikaplun@etu.ru (D.K.); k-fmf-primath@stavsu.ru (P.L.)
2. Scientific Research Institute of System Analysis, Russian Academy of Sciences, 117218 Moscow, Russia; aserg@cs.niisi.ras.ru
3. Military Academy after Peter the Great, 142210 Serpuhov, Russia; aveligosha@mail.ru
4. Laboratory of Computer Security Problems, Saint Petersburg Institute for Informatics and Automation of the Russian Academy of Sciences, 199178 Saint Petersburg, Russia; doynikova@comsec.spb.ru
5. Department of Applied Mathematics and Mathematical Modeling, North-Caucasus Federal University, 355009Stavropol, Russia
6. Youth Research Institute, Saint Petersburg Electrotechnical University "LETI", 197022 St. Petersburg, Russia
* Correspondence: dnbutusov@etu.ru

Received: 18 November 2019; Accepted: 17 December 2019; Published: 19 December 2019

Featured Application: Nowadays digital filters are widely used in receivers of different software-defined radio (SDR) communication systems. The main factor affecting the development of SDR is the characteristics of analog-to-digital and digital-to-analog converters. SDR technology allows us to replace existing and developed designs of receivers and transceivers of a heterodyne structure with a limited number of hardware units controlled by specialized software. This helps to simplify the constructions, make them cheaper, improve their performance, and support any modulation types. Furthermore, such an approach can be fruitful in signal reception and demodulation for different types of digital modulations such as DPSK, QAM, GMSK, etc. The main operations in the SDR receiver are heterodyning and filtering, which are performed digitally. In this case, digital filtering determines almost all parameters of the output channel of such a receiver. Therefore, the parameters of digital filters in these receivers have to meet more strict requirements, including the accuracy of calculations and hardware costs. The use of finite field algebra will significantly increase the accuracy of calculations in digital filters of such receivers and reduce hardware costs.

Abstract: The applications of digital filters based on finite field algebra codes require their conjugation with positional computing structures. Here arises the task of algorithms and structures developed for converting the positional notation codes to finite field algebra codes. The paper proposes a method for codes conversion that possesses several advantages over existing methods. The possibilities and benefits of optimization of the computational channel structure for digital filter functioning based on the codes of finite field algebra are shown. The modified structure of computational channel is introduced. It differs from the traditional structure by the fact that there is no explicit code converter in it. The main principle is that the "reference" values of input samples, which are free from the error of the analog-digital converter, are used as input samples. The proposed approach allows achieving a higher quality of signal processing in advanced digital filters.

Keywords: digital filter; finite field algebra; conversion device; module; memory device; residue

1. Introduction

For the effective implementation of digital signal processing (DSP) algorithms, especially digital filters (DF), number-theoretic methods based on prime numbers [1–3] are of great importance. Many of these methods allow parallel computing, thus the research on theory and application of numerical systems with parallel structure is of particular interest in the field of DSP, image processing systems, cryptographic systems, quantum automated machines, neural computers systems, massive concurrency of operations, cloud computing, etc. [4–9]. Such systems are best suited for parallel computing. One of the most fruitful research areas here is the algebra of finite field, which provides an impressive level of internal parallelism [10] to the DSP systems.

Recent studies in designing DSP computing devices based on the finite field algebra (FFA) have shown that FFA has a superior potential for improvement of performance and reliability of numerical information processing being compared with the traditional positional numeral system (PNS) [11].

Since the FFA is an integer numeral system, it is possible to represent the processed data in DSP devices with arbitrary accuracy. From [11,12], it is known that the FFA advantages appear most clearly when tabular schemes are in use. Therefore, as the integral technology improves (e.g., the production of storage devices with high information density) along with the technical basis of the tabular computational method, efficiency of using FFA codes is steadily increasing.

Most popular designs of computing devices operating in FFA codes are focused on the implementation of computational processes of the same type despite their different specialization. These processes are the sequences of arithmetic operations (addition and multiplication) with integer numbers. This determines interest in researching the FFA implementation in highly efficient digital filtering algorithms. The following advantages of FFA can be outlined from [11–13]:

1. Independence of formation of numbers' bits. Whereby each bit carries information about the entire original number instead of the intermediate number resulting from the formation of lower-order bits. This implies the possibility of numbers' bits independent parallel processing.

2. The low bitness of residues representing a number. It results from a small number of possible code combinations. It allows using tabular arithmetic where the typical arithmetic operations are transformed into operations performed by simple selection of the result of calculations from the table (memory device).

3. The FFA has natural corrective abilities. The FFA codes allow efficient detection and correction of errors while transmitting signals and performing arithmetic operations [14].

Considering the abovementioned issues, we can conclude that it is advisable to use the FFA for the synthesis of DF with the required quality indicators. The FFA advantages compared with PNS allow providing the required frequency and accuracy characteristics of filters and digital signal processing in real-time [13]. Modern means of digital signal processing (for example, digital receivers) have strict requirements for the quality of signal processing. The analysis carried out in [12,13] showed that the use of FFA allows one to achieve the required indicators of signal processing quality. Let us give some preliminaries first.

2. The Basics of Operations on Numbers in the FFA

The theoretical basis of FFA is the theory of comparisons. Two integers, A_1 and A_2, that have the same residues being divided by module p are called comparable in modulus p, and the relation of comparability takes the following form:

$$A_1 \equiv A_2 \bmod p \qquad (1)$$

From view of numbers comparability only residue α is used. It is obtained by dividing the number A by the number p. Thus, the following comparison is true:

$$\alpha \equiv A \bmod p \qquad (2)$$

Finding the residue is the transformation of the number A modulo p. The operation of determining the residue is performed by the following rule [11]:

$$\forall A \in Z : |A|_{p^+} \leftrightarrow A - [A/p]_p \tag{3}$$

The residues of number A modulo p will belong to the number range $\alpha \in (0, 1, 2, \ldots, p-1)$. While performing calculations it is always possible to replace the comparison that establishes a relationship between integer classes of numbers having the same residue with the equality including this residue. For example, if

$$A + B \equiv C \bmod p \tag{4}$$

Than Equation (4) can be written as follows:

$$(A + B) \bmod p = C \bmod p \tag{5}$$

Calculations with residues are rather simple since they can get values no more than $p-1$ [11,12]. Therefore

$$\begin{array}{l}(A \pm B) \bmod p \text{ is equivalent to } (A \bmod p + B \bmod p) \bmod p \\ A \cdot B \bmod p \text{ is equivalent to } (A \bmod p \cdot B \bmod p) \bmod p\end{array} \tag{6}$$

Therefore, for any operations of multiplication, addition, subtraction, one can replace the result of calculations at each step by its residue. Representation of numbers in the FFA is provided by the smallest non-negative residues α_i according to the system of mutually simple modules in the following form:

$$p_i(\forall i \in [1; n]; A(\alpha_1, \alpha_2, \ldots, \alpha_n)). \tag{7}$$

Addition, subtraction, and multiplication of two numbers A and B can be performed by the addition, subtraction, or multiplication of the residues α_i and β_i for each module p_i, independently. If the value P is chosen as the product of modules p_i, then actions with large numbers can be performed in such a system with a large number of small modules p_i. The value P determines the complete range of representation of numbers in the FFA code.

The following identity can be written [11]:

$$\forall A \in (0, 1, 2, \ldots, p_i - 1) \bmod p_i : A = \left| \sum_{n-1}^{n} |A|_{pi}^+ + m_i P_{in} \right|_{P_n} \tag{8}$$

where $m_i = \left| P_{in}^{-1} \right|_{pi}^+$.

The identity (8) is the basis for generating the finite field arithmetic code. If the fixed series of positive integers $p_1, p_2, \ldots, p_n$ are modules than the finite field arithmetic (the system of residual classes) is a such nonpositional number system in which any positive integer is represented as a set of residues from dividing the represented number by the selected base of the system as follows:

$$A(\alpha_1, \alpha_2, \ldots, \alpha_n) \tag{9}$$

where $\alpha_1, \alpha_2, \ldots, \alpha_n$—the smallest non-negative number residues by modules $p_1, p_2 \ldots, p_n$, respectively.

The numbers α_i by the selected modules are formed as follows:

$$\alpha_i = \text{rest } A \ (\bmod \ p_i) = A - \left[\frac{A}{P_i}\right] p_i \ (\forall i \in [1, n]) \tag{10}$$

where A/p_i—integer quotient; p_i—bases-mutually prime numbers.

In number theory, it is proved [11,12] that if "$i \neq j$, $(p_j, p_i) = 1$", then the number representation (10) is the only one if $0 \leq A \leq P_n$, where $P_n = p_1 \cdot p_2 \cdot \ldots \cdot p_n$—is the number range, i.e., there is the number A for which

$$\begin{cases} A \equiv \alpha_1 \pmod{p_1} \\ A \equiv \alpha_2 \pmod{p_2} \\ \ldots \\ A \equiv \alpha_n \pmod{p_n} \end{cases} \quad (11)$$

Thus, the conclusion can be done that is advisable to use the methods of organizing calculations based on the representation of the processed data in the FFA codes in the algorithms of digital filtering. It should be noticed that this task is solved on the basis of a systematic approach. Namely, for the synthesis of DF operating on the basis of the FFA codes a number of tasks should be solved including:

- the task of converting the processed data from the positional representation to the FFA code. It requires the development of an efficient way of converting;
- the task of converting the processed data represented in the FFA code to the positional representation. It requires the development of an efficient way of converting;
- the task of implementation of the digital filtering algorithm in the FFA codes. The implemented algorithm should provide the required quality indicators of the filter output signal (accuracy, the calculation time of the output sample, reliability);
- the task of providing the required degree of structural fault tolerance of the filter [15].

Currently, there are numerous studies devoted to solving the abovementioned tasks and other problems in DSP. Their solution will allow one to fully utilize the advantages of the FFA and ensure efficient signal processing in the digital filters [12,13]. In Figure 1, the variant of a simplified structural diagram of DF operating in the FFA codes is represented.

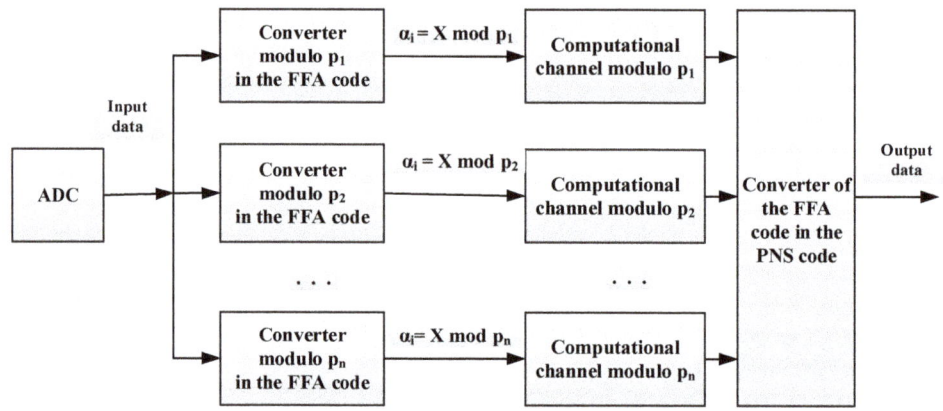

Figure 1. The variant of simplified structural diagram of digital filters (DF) operating in the finite field algebra (FFA) codes.

3. Digital Filters in the FFA

Here we describe an implementation of the converter of position code to the FFA code. An opportunity to exclude an influence of the errors of analog-to-digital converter on the filter output sample is considered.

Analysis of the modern implementations of computational algorithms in the FFA codes allows concluding that the time of reverse conversion to positional representation in them takes more than 50% of the common time of calculations [11,12].

Insufficient development of the theoretical foundations of construction of code converters and, following this, the insufficient development of methods and means for their implementation becomes a critical place in the entire cycle of development and implementation of computing devices of DF operating on the basis of the FFA. It leads to loss of its advantages.

In [13,16,17] the algorithms of operation and hardware implementation of the devices for interfacing the positioning and computing FFA devices are reviewed. The devices in which the determination of residue is based on the use of the property of residues' cyclical nature are considered. Following the Equation (1)

$$|\alpha_1|^+_{p_n} \in \{0 \ldots p_n - 1\}, \tag{12}$$

the residue α_i will repeat d times in the range of convertible numbers. The cycle of its reiteration will depend on the value of module p_i. In other words, the value d can be determined from the Equations (13) and (14).

$$d =]\frac{2^s}{p_i}[, \text{ for } \alpha_i \in \{0, \ldots, p_i - 1\} \tag{13}$$

$$d =]\frac{2^s}{p_i}[+ 1, \text{ for } \alpha_i \in \{0, \ldots, g\} \tag{14}$$

where g—is the residue of dividing the number 2^s by p_i; s—is the bit width of the number being converted in the FFA code. In accordance with the Equations (13) and (14) the residue α_i corresponds to d numbers from the range of 2^s. To calculate the residue α_i of initial number A it should be uniquely determined that the number $A \in \{d\}$, i.e.,

$$A \in d \text{ for } d =]\frac{2^s}{p_i}[, \text{ for } \alpha_i \in \{0, \ldots, p_i - 1\}$$

$$A \in d \text{ for } d =]\frac{2^s}{p_i}[+ 1, \text{ for } \alpha_i \in \{0, \ldots, g\}$$

For this goal the range of binary numbers presented in the FFA code can be divided into the subranges. The number of subranges and the values of numbers in them are determined by the value $E = 2^{s/2}$. The values of numbers in subranges will be within the numbers' intervals with a step equal to one. They will be determined by the following expressions:

$$\begin{aligned} E_1 &= 0, 1, \ldots, 2^{s/2} - 1; \\ E_2 &= 2^{s/2}, 2^{s/2} + 1, \ldots, 2^{\frac{s}{2}+1} - 1; \\ &\ldots \\ E_k &= 2^{(\frac{s}{2})+(\frac{s}{2}-1)}, 2^{(\frac{s}{2})+(\frac{s}{2}-1)} + 1, \ldots, \quad 2^s - 1. \end{aligned} \tag{15}$$

From the expression (15) it follows that the upper $s/2$ bits of number unambiguously determine the number of subrange E_i in which the number is. While the lower $s/2$ bits of number A determines the index of number in the subrange. Thus, based on the cyclicity property of residues modulo, the finding of residue for the number A will include determination of subrange number and reading the residue from the memory device for each subrange. Considering the Equations (13)–(15), the algorithm of residue finding on the basis of subrange determination will include the following procedures:

1. Determining the numbers subrange in the positional numeral system (PNS) $D_p = 2^s$
2. Splitting the range D_p into subranges (15).
3. Determining the subrange by the $s/2$ upper bits of initial number A.
4. Getting the residue α_i from the memory device using the specific address by the $s/2$ lower bits of initial number A.

The block diagram of the proposed interfacing device includes the register where the initial number A is written, the $2^{s/2}$ comparison schemes and the $2^{s/2}$ memory devices. The proposed method of residue finding provides it for two modular cycles of the converter. The time of residue formation does not change with an increase in the bit width of the converted source numbers.

The disadvantage of the proposed algorithm is that the hardware costs required for its implementation are high. In addition, the residue values repeat in the memory devices. This indicates an incomplete use of modular code ring properties.

While researching the proposed algorithm of formation of a modulo residue it was found that the first values of residues in the subranges of expression (15) for the modules getting the values $p_i < 2^{s/2}$ change by a value

$$C_s = 2^{s/2} - p_i, \qquad (16)$$

while for the modules getting the values $p_i > 2^{s/2}$, these values are constant and determined as

$$C_s = 2^{s/2}. \qquad (17)$$

Using these properties, the algorithm of residue determination can be described as follows:

$$\alpha_i = |R + C_s|, \qquad (18)$$

where R—the number determined by the $s/2$ lower bits of the number A; C_s—the value calculated using Equation (16) and written to the memory device (MD) by the address that gets the value in accordance with the upper $s/2$ bits of number A. The block diagram of the proposed converting device is provided in Figure 2.

The developed algorithm provides the residue formation for three modular cycles of the interfacing device. Thus, we can conclude that the developed conversion method provides significantly better performance than the existing conversion methods considered in [10].

It should be noted that the algorithm provides operation in the conveyor mode, i.e., matching the speed of arrival of input data into the computing device of a DF and calculation of its output sample on the basis of FFA.

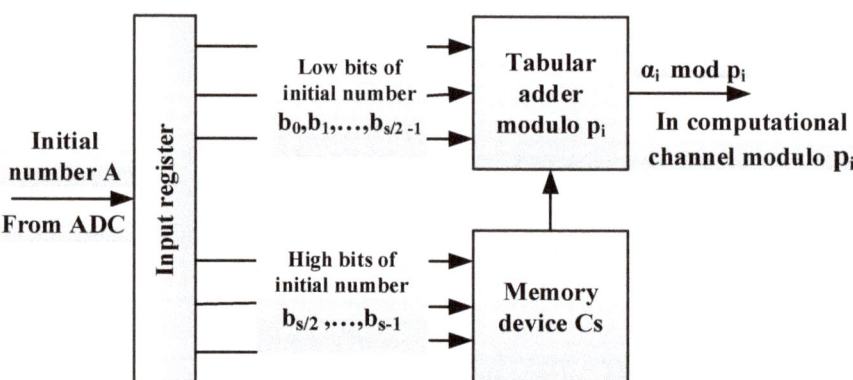

Figure 2. The block diagram of the proposed converting device.

The abovementioned method of data representation in the FFA codes meets the requirements of the real-time signal processing devices in terms of performance indicators. Application of the conversion device for data in the FFA codes complicates the overall filter structure, requires additional costs to synchronize the operation of its elements in the mode of calculations conveyor. The question arises about the possibility of excluding the conversion device from the structure of the computation channel of DF to improve the accuracy of the calculation of the output sample.

In [15,16] the models of calculation accuracy for output samples in a positional digital filter and in the filter operating on the basis of FFA are provided. The model of calculation accuracy in the positional DF considering that while calculating the output sample the intermediate results will be rounded off, can be represented as follows:

$$e_{er\ com}(nT) = Q_{ADC}(nT) + e_{q\ coef}(nT) + e_{q\ is}(nT) + e_{ro\ out}(nT) + e_{add}(nT), \tag{19}$$

where $e_{er\ com(nT)}$—the common calculation error for the output sample; $Q_{ADC}(nT)$—the error of analog-to-digital converter (ADC); $e_{q\ coef}(nT)$—the error of coefficients quantization when they are represented in a computational digital filtering algorithm; $e_{q\ is}(nT)$—the error of input sample quantization when they are represented in a computational digital filtering algorithm; $e_{ro\ out}(nT)$—the error introduced by the intermediate results rounding off; $e_{add}(nT)$—the error resulting from the fact that the input of each subsequent stage will receive an intermediate sample, which already has an error that accumulates when the intermediate sample "passes" through the stages of the filter.

Moreover, the common model of calculation accuracy for an output sample in the positioning DF considering that the intermediate results will be truncated can be represented as follows:

$$e_{er\ com}(nT) = Q_{ADC}(nT) + e_{q\ coef}(nT) + e_{q\ is}(nT) + e_{tr\ out}(nT) + e_{add}(nT) \tag{20}$$

where $e_{tr\ out}(nT)$—the error introduced by truncation of intermediate calculation results in the filter links.

In the computational device of DF, operating on the basis of FFA, there is no such disadvantages as the operations of truncation (rounding off) of intermediate calculation results, the additional errors, the errors of quantification of input data and filter coefficients. Therefore, there is no accumulation of errors in the filter when calculating the output sample.

Then the accuracy model of output sample calculation for the DF operating on the basis of FFA can be represented as follows [14]:

$$e_{er\ com}(nT) = Q_{ADC}(nT). \tag{21}$$

Analysis of Equations (19)–(21) allows concluding that the accuracy of DF operating on the basis of FFA is significantly higher than the accuracy of positioning filters. Analysis of the influence of errors that occur when calculating the output sample of the FFA filter during signal processing in radio channels showed that the signal-to-noise ratio (SNR) is about 68–70 dB.

If the amplitude of the input signal is less than a half of the voltage of the full scale of ADC, then there is an additional attenuation about −20 dB. In this case, the SNR value is calculated as follows: SNR = 74 dB − 20 dB = 54 dB. However, as soon as radio channels are subject to significant fading, it can be argued that the SNR value will be 40–50 dB.

Thus, it can be concluded that in case of exclusion of ADC from the signal processing path, for example, from the digital receiver path (Figures 3 and 4) the value of SNR can be increased.

As soon as while building the digital receiver the range of data being processed in the digital signal processing device (Figures 3 and 4) is known, namely, the bit width of used ADC is known, then there is an opportunity to represent input data being processed in the filters' computational channel in the FFA codes without ADC and the device of their conversion. Considering Equation (12), the values of residues for any module of the selected base system do not exceed the value of the module and their count is equal to the module value. Then as a source of input data for the DF's computational channel a memory device can be used. It records the values of residuals for the selected module.

Therefore, the FFA ring property allows one to significantly simplify the hardware implementation of the computing device of the filter, to increase its performance and to exclude the error of output sample calculation. Thus

$$e_{er\ com}(nT) = 0. \tag{22}$$

In the case of such calculations, the error introduced by analog-to-digital conversion of input data is excluded from the channel of FFA digital filter. This further increases the accuracy of the calculation of the filter output sample. The structure of the computational channel of DF operating based on the FFA is shown in Figure 5.

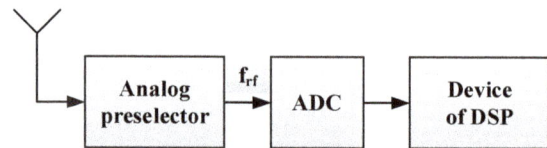

Figure 3. The structure of digital receiver by radio frequency.

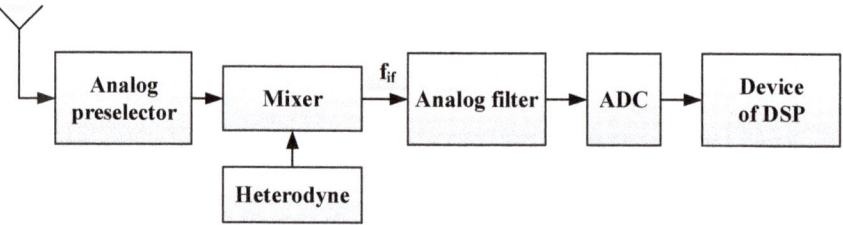

Figure 4. The structure of digital receiver by intermediate frequency.

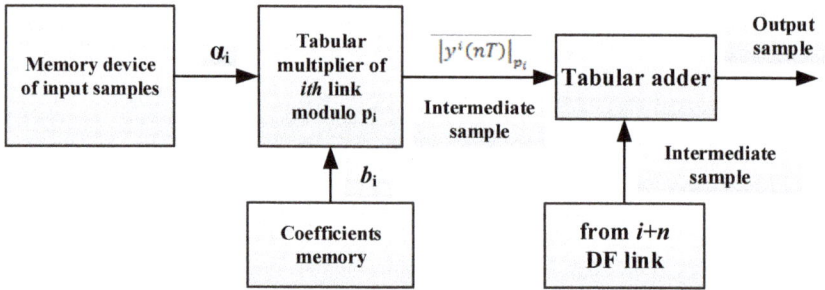

Figure 5. The structure of the computational channel of DF operating based on the FFA.

Thus, the data converter from positional representation to the FFA code can be excluded from the DF structure. The ADC can be used as a control device for the selection of the required residue from the memory device of the output sample. In this case, one ADC can be used for all computational channels of the FFA DF.

4. Testing of DF in the FFA

Let us evaluate the proposed solutions experimentally. Evaluation of the efficiency of the proposed conversion method is conducted based on the comparison of its conversion time with the existing methods. Evaluation is made considering the number of operation cycles of comparable devices. It is provided in Table 1.

Table 1. Comparative evaluation of performance of existing and developed conversion methods.

Conversion Method		Number of Operation Cycles of Conversion Device		Benefit in Performance, Times (Of the Developed Method to the Existing Methods)
		Bit Width of Initial Data, Bit		
		16	32	
Lowering bit width [18]	Serial adder	19	31	6.3/10.3
	Parallel adder	9	9	3/3
The method based on the block adders residue [19]		12	18	4/6
The method based on the hierarchical neural network of finite ring [20]		24	36	8/12
The method based the distributed arithmetic and the parallel neural network [20]		15	15	5/5
The enhanced method based on the distributed arithmetic and the parallel neural network [20]		11	11	3.7/3.7
The developed method based on the determination of subranges		3	3	-

The data provided in Table 1 allows concluding that the proposed method for data conversion is efficient and can be used to build DF operating in the FFA codes.

The existing methods for data conversion from the positional representation to the FFA code are based on the sequential bitwise conversion of the source number. To get the modulo residue the arithmetic operations with the number bits are conducted. The type and the number of arithmetic operations depend on the conversion method. It determines the number of operation cycles for the conversion device (see Table 1).

Let the base system be given $p_1 = 5$, $p_2 = 9$, $p_3 = 13$. The range of the processed data in this case is $p = 585$. In the case of this implementation, the DF structure will include three computational channels (see Figure 1). Each computational channel, except the computational device implementing the filtering algorithm, includes the data conversion device (see Figure 2).

In the considered case, when the conversion device is excluded from the structure of the calculation (Figure 5), in the memory device for the input sample, for example, modulo $p_2 = 9$, the residue values for $p = 585$ will be recorded in the form of Table 2.

Table 2. The residues values for the range $p = 585$ modulo $p_2 = 9$.

The Residues Values	The Range Numbers	The Residues Values	The Range Numbers
$a_1 = 0$	0,9,18,27,36,45,54,63,72,81, ...	$a_5 = 4$	4,13,22,31,40,49,58,67,76,85, ...
$a_2 = 1$	1,10,19,28,37, 46,55,64,73,82, ...	$a_6 = 5$	5,14,23,32,41,50,59,68,77,86, ...
$a_3 = 2$	2,11,20,29,38, 47,56,65,74,83, ...	$a_7 = 6$	6,15,24,33,42,51,60,69,78,87, ...
$a_4 = 3$	3,12,21,30,39, 48,57,66,75,84, ...	$a_8 = 7$	7,16,25,34,43,52,61,70,79,88, ...
		$a_9 = 8$	8,17,26,35,44,53,62,71,80,89, ...

Let us consider an example of a calculation of residue value for the number $X = 32{,}015$ modulo $p = 17$ for the method "Lowering the bit width with the parallel adder" (Table 1). Representing the number $X = 32{,}015$ in binary code and dividing it into blocks of 4 bits we obtain

$$X = 32015 = 0111\ 1101\ 0000\ 1111,$$

i.e., we have 4 numbers $a_3 = 7$, $a_2 = 13$, $a_1 = 0$, $a_0 = 15$. Then we can write

$$|32015|_{17}^+ = \|7 \cdot 2^{12}|_{17}^+ + |13 \cdot 2^8|_{17}^+ + |0 \cdot 2^4|_{17}^+ + |15 \cdot 2^0|_{17}^+|_{17}^+ =$$
$$= \|7 \cdot 16|_{17}^+ + |13 \cdot 1|_{17}^+ + |0 \cdot 16|_{17}^+ + |15 \cdot 1|_{17}^+|_{17}^+ = |10 + 13 + 0 + 15|_{17}^+ = 4$$

In the case of parallel implementation of this algorithm, the time of residue calculation is

$$T_{FC} = t_{LUT} + t_{LUT} \times log_2 \left\lceil \frac{n}{k} \right\rceil \tag{23}$$

where k—the block size (in the example $k = 4$); $n = 16$—the bit width of the input number X; t_{LUT}—time of getting from the LUT-table (is taken equal to three cycles).

For the considered example, the time of residue calculation considering Equation (23) is equal to nine cycles, which corresponds to the Table 1.

To implement the considered converter (the number of hardware devices is brought to the amount of LUT-tables) eleven LUT-tables are required. For the suggested method (Figure 2), three LUT-tables are required. For the computational channel without conversion device (Figure 5), one LUT-table is required.

Following Equation (21), the error introduced to the processes signal by the twelve bit ADC is Δ = 0.00024414. As an example, a non-recursive DF of the forty-fifth order is taken [13]. For evaluation the samples of impulse response of the filter taking values $N_1 = -0.000105023$, $N_2 = -0.000125856$, $N_{17} = 0.0364568$, $N_{18} = 0.0328505$ are selected. To simplify the provided values the sample sign of the impulse response is not considered and it is supposed that the ADC error decreases the values of the selected samples.

While calculating the output sample of the filter and representing the samples of impulse response in the binary code considering the ADC errors, their values can be written as follows: $N_1 = 0.000139117$, $N_2 = 0.0001185284$, $N_{17} = 0.03621266$, $N_{18} = 0.03260636$.

The values of samples of impulse response in the FFA code for the module $p_1 = 5$ without error (in decimal notation and in FFA code) will be written as follows:

$N_1 = (105023)_{10} = (3)_{FFA}$,
$N_2 = (125856)_{10} = (1)_{FFA}$,
$N_{17} = (36456800)_{10} = (0)_{FFA}$,
$N_{18} = (32850500)_{10} = (0)_{FFA}$.

Considering the error introduced by the ADC, the values of samples of impulse response in the FFA code for the module $p_1 = 5$ will be written (in decimal notation and in FFA code) as follows:

$N_1 = (139117)_{10} = (2)_{FFA}$,
$N_2 = (1185284)_{10} = (4)_{FFA}$,
$N_{17} = (362126600)_{10} = (0)_{FFA}$,
$N_{18} = (326063600)_{10} = (0)_{FFA}$.

As it follows from the conducted evaluation the values of the first and the second samples in the FFA code changed because of the ADC error influence. This error will change the filter frequency response in the process of calculation of its output sample. Considering that the error introduced by the ADC can affect the values of the significant number of impulse response samples, the distortion of frequency response will be significant.

5. Discussion

The known methods [18–21] for data conversion from the positional representation to the FFA code are based on the sequential bitwise conversion of the source number. To get the modulo residue the arithmetic operations with the number bits are conducted. The type and the number of arithmetic operations depend on the conversion method. It determines the number of operation cycles for the conversion device. Our study has shown that the data converter from positional representation to the FFA code can be excluded from the DF structure. In this case, the ADC can be used as a control device for the selection of the required residue from the memory device of the output sample. We described the structure of the computational channel and provided several practical tests, aimed both at a speed characteristics study and error estimation. The optimization technique was given and experimentally tested.

The obtained results open the possibility for efficient and compact hardware implementation of the digital filters on modern devices (DSP, FPGA, ASIC, etc.) for processing signals in various areas, such as radio communication, hydroacoustics, radars and echolocation systems, as well as in industry, defense, law enforcement, and other fields of science and technology [22]. Further research will be devoted to the comparison of the proposed approach with existing techniques of low-precision digital filters [23] and control systems [24] implementation based on an alternative discrete operator technique [25,26] and Gaussian approximation approach [27]. We will also try to build efficient adaptive DF [28] based on the proposed approach.

6. Conclusions

Thus, the conclusion can be made that building of computational channel of the DF operating in the FFA codes without the converter and ADC influence on the input data values allows increasing accuracy of calculation of filter output sample. Similar estimates are valid not only for the impulse response samples of the filter but also for the input signals. In the first case, the impulse response of the filter is changed. In the second case, the output sample will be distorted. For example, if an adaptive DF will be implemented, the error of the output sample will require setting up the filter coefficients. It, in turn, will require high time costs.

The actual variant of construction and organization of calculations in the computational channel of the DF will depend on the requirements for performance and accuracy of the output sample calculation.

Finally, we can conclude that the properties of finite field algebra ensure the construction of efficient computational algorithms and structures of DF with high performance and accuracy of the output sample calculation.

Author Contributions: Conceptualization, D.K. and S.A.; bata curation, A.V. and D.B.; formal analysis, E.D. and D.B.; funding acquisition, D.K.; investigation, S.A. and D.B.; methodology, A.V. and P.L.; project administration, D.K., S.A., and P.L.; resources, S.A. and E.D.; software, A.V. and P.L.; supervision, D.K. and S.A.; validation, P.L. and D.B.; visualization, A.V. and E.D.; writing—original draft, D.K., A.V., and D.B.; writing—review and editing, E.D. and D.B. All authors have read and agreed to the published version of the manuscript.

Funding: This research was funded by the grant of the Russian Science Foundation (Project №19-19-00566).

Conflicts of Interest: The authors declare no conflict of interest.

References

1. Machado, J.T.; Lopes, A.M. Multidimensional Scaling and Visualization of Patterns in Prime Numbers. *Commun. Nonlinear Sci. Numer. Simul.* **2020**, *83*, 105128. [CrossRef]
2. Guariglia, E. Primality, fractality, and image analysis. *Entropy* **2019**, *21*, 304. [CrossRef]
3. Zhang, Y. Bounded gaps between primes. *Ann. Math.* **2014**, *179*, 1121–1174. [CrossRef]
4. Szabo, N.S.; Tanaka, R.I. *Residue Arithmetic and Its Applications to Computer Technology*; McGraw-Hill: New York, NY, USA, 1967.
5. Molahosseini, A.S.; Sorouri, S.; Zarandi, A.A.E. Research challenges in next-generation residue number system architectures. In Proceedings of the 7th International Conference on Computer Science & Education (ICCSE), Melbourne, VIC, Australia, 14–17 July 2012; pp. 1658–1661. [CrossRef]
6. Mohan, P.V.A. *Residue Number Systems: Theory and Applications*; Birkhäuser: Basel, Switzerland, 2016.

7. Chervyakov, N.I.; Lyakhov, P.A.; Nagornov, N.N.; Kaplun, D.I.; Voznesenskiy, A.S.; Bogayevskiy, D.V. Implementation of Smoothing Image Filtering in the Residue Number System. In Proceedings of the 8th Mediterranean Conference on Embedded Computing (MECO), Budva, Montenegro, 10–14 June 2019. [CrossRef]
8. Younes, D.; Steffan, P. A comparative study on different moduli sets in residue number system. In Proceedings of the International Conference on Computer Systems and Industrial Informatics (ICCSII), Sharjah, United Arab Emirates, 18–20 December 2012; pp. 1–6. [CrossRef]
9. Nakahara, H.; Nakanishi, H.; Iwai, K.; Sasao, T. An FFT circuit for a spectrometer of a radio telescope using the nested RNS including the constant division. *ACM SIGARCH Comput. Archit. News* **2017**, *44*, 44–49. [CrossRef]
10. Kalmykov, I.A.; Veligosha, A.V.; Kaplun, D.I.; Klionskiy, D.M.; Gulvanskiy, V.V. Parallel-pipeline implementation of digital signal processing techniques based on modular codes. In Proceedings of the XIX conference on Soft Computing and Measurements (SCM), St. Petersburg, Russia, 25–27 May 2016; pp. 213–214.
11. Kaplun, D.; Butusov, D.; Ostrovskii, V.; Veligosha, A.; Gulvanskii, V. Optimization of the FIR Filter Structure in Finite Residue Field Algebra. *Electronics* **2018**, *7*, 372. [CrossRef]
12. Omondi, A.; Premkumar, B. *Residue Number Systems: Theory and Implementation*; Imperial College Press: London, UK, 2007.
13. Veligosha, A.V.; Kaplun, D.I.; Gulvanskiy, V.V.; Klionskiy, D.M.; Kupriyanov, M.S. Implementation of digital filters in the residue number system. In Proceedings of the IEEE NW Russia Young Researchers in Electrical and Elec-tronig Engineering Conference (EIConRusNW), St. Petersburg, Russia, 2–3 February 2016; pp. 220–224.
14. Veligosha, A.V.; Kaplun, D.I.; Klionskiy, D.M.; Bogaevskiy, D.V.; Gulvanskiy, V.V.; Kalmykov, I.A. Error Correction of Digital Signal Processing Devices using Non-Positional Modular Codes. *Autom. Control Comput. Sci.* **2017**, *51*, 167–173.
15. Veligosha, A.; Kaplun, D.; Voznesenskiy, A.; Bogaevskiy, D. Structural and informational diversity of digital filters based on multivariate arithmetic of finite field. In Proceedings of the Conference of Open Innovation Association (FRUCT), Moscow, Russia, 10–12 April 2019; pp. 479–485. [CrossRef]
16. Kaplun, D.I.; Gulvanskiy, V.V.; Klionskiy, D.M.; Kupriyanov, M.S.; Veligosha, A.V. Implementation of non-positional digital filters. In Proceedings of the XIX IEEE International Conference on Soft Computing and Measurements (SCM), St. Petersburg, Russia, 25–27 May 2016; pp. 220–224.
17. Veligosha, A.V.; Bratchenko, N.Y.; Kaplun, D.I.; Klionskiy, D.M.; Gulvanskiy, V.V.; Bogaevskiy, D.V. Data representation in the modular code. In Proceedings of the 38th Progress in Electromagnetics Research Symposium, St. Petersburg, Russia, 22–25 May 2017; pp. 449–452. [CrossRef]
18. Chervyakov, N.I.; Lyakhov, P.A.; Babenko, M.G.; Lavrinenko, I.N.; Lavrinenko, A.V.; Nazarov, A.S. The architecture of a fault-tolerant modular neurocomputer based on modular number projections. *Neurocomputing* **2018**, *272*, 96–107. [CrossRef]
19. Zhang, D.; Jullien, G.A.; Miller, W.C. A neural-like network approach to finite ring computations. *IEEE Trans. Circuits Syst.* **1990**, *37*, 1048–1052. [CrossRef]
20. Chervyakov, N.I.; Lyakhov, P.A.; Babenko, M.G.; Garyanina, A.I.; Lavrinenko, I.N.; Lavrinenko, A.V.; Deryabin, M.A. An efficient method of error correction in fault-tolerant modular neurocomputers. *Neurocomputing* **2016**, *205*, 32–44. [CrossRef]
21. Mitra, S.K. *Digital Signal Processing: A Computer-Based Approach*; McGraw-Hill: New York, NY, USA, 1998.
22. Molahosseini, A.; Sousa, L.D.; Chang, C. *Embedded Systems Design with Special Arithmetic and Number Systems*; Springer: Cham, Switzerland, 2017.
23. Karimov, T.I.; Butusov, D.N.; Andreev, V.S.; Rybin, V.G.; Kaplun, D.I. Compact Fixed-Point Filter Implementation. In Proceedings of the 22nd Conference of FRUCT Association, Jyvaskyla, Finland, 15–18 May 2018; Volume 2, pp. 73–78.
24. Middleton, R.H.; Goodwin, G.C. Improved Finite Word Length Characteristics in Digital Control Using Delta Operators. *IEEE Trans. Autom. Control* **1986**, *31*, 1015–1021. [CrossRef]
25. Butusov, D.N.; Karimov, T.I.; Kaplun, D.I.; Karimov, A.I. Delta operator filter design for hydroacoustic tasks. In Proceedings of the 6th Mediterranean Conference on Embedded Computing (MECO), Bar, Montenegro, 11–15 June 2017; pp. 1–4.

26. Kauraniemi, J.; Laakso, T.I.; Hartimo, I. Delta Operator Realizations of Direct-Form IIR Filters. *IEEE Trans. Circuits Syst. II Analog Digit. Signal Process.* **1998**, *45*, 41–52. [CrossRef]
27. Capizzi, G.; Coco, S.; Sciuto, G.L.; Napoli, C. A New Iterative FIR Filter Design Approach Using a Gaussian Approximation. *IEEE Signal Process. Lett.* **2018**, *25*, 1615–1619. [CrossRef]
28. Veligosha, A.V.; Kaplun, D.I.; Bogaevskiy, D.V.; Gulvanskiy, V.V.; Voznesenskiy, A.S.; Kalmykov, I.A. Adjustment of adaptive digital filter coefficients in modular codes. In Proceedings of the IEEE North West Russia Section Young Researchers in Electrical and Electronic Engineering Conference (EIConRusNW), St. Petersburg, Russia, 29 January–1 February 2018; pp. 1167–1170.

© 2019 by the authors. Licensee MDPI, Basel, Switzerland. This article is an open access article distributed under the terms and conditions of the Creative Commons Attribution (CC BY) license (http://creativecommons.org/licenses/by/4.0/).

Article

Multiresolution Speech Enhancement Based on Proposed Circular Nested Microphone Array in Combination with Sub-Band Affine Projection Algorithm

Ali Dehghan Firoozabadi [1,*], Pablo Irarrazaval [2,3,4], Pablo Adasme [5], David Zabala-Blanco [6,*], Hugo Durney [1], Miguel Sanhueza [1], Pablo Palacios-Játiva [7] and Cesar Azurdia-Meza [7]

1. Department of Electricity, Universidad Tecnológica Metropolitana, Av. José Pedro Alessandri 1242, Santiago 7800002, Chile; hdurney@utem.cl (H.D.); msanhueza@utem.cl (M.S.)
2. Electrical Engineering Department, Pontificia Universidad Católica de Chile, Santiago 7820436, Chile; pim@uc.cl
3. Biomedical Imaging Center, Pontificia Universidad Católica de Chile, Santiago 7820436, Chile
4. Institute for Biological and Medical Engineering, Pontificia Universidad Católica de Chile, Santiago 7820436, Chile
5. Electrical Engineering Department, Universidad de Santiago de Chile, Av. Ecuador 3519, Santiago 9170124, Chile; pablo.adasme@usach.cl
6. Department of Computing and Industries, Universidad Católica del Maule, Talca 3466706, Chile
7. Department of Electrical Engineering, Universidad de Chile, Santiago 8370451, Chile; pablo.palacios@ug.uchile.cl (P.P.-J.); cazurdia@ing.uchile.cl (C.A.-M.)
* Correspondence: adehghanfirouzabadi@utem.cl (A.D.F.); dzabala@ucm.cl (D.Z.-B.); Tel.: +56-2-2787-7117 (A.D.F.)

Received: 6 April 2020; Accepted: 4 June 2020; Published: 6 June 2020

Abstract: Speech enhancement is one of the most important fields in audio and speech signal processing. The speech enhancement methods are divided into the single and multi-channel algorithms. The multi-channel methods increase the speech enhancement performance by providing more information with the use of more microphones. In addition, spatial aliasing is one of the destructive factors in speech enhancement strategies. In this article, we first propose a uniform circular nested microphone array (CNMA) for data recording. The microphone array increases the accuracy of the speech processing methods by increasing the information. Moreover, the proposed nested structure eliminates the spatial aliasing between microphone signals. The circular shape in the proposed nested microphone array implements the speech enhancement algorithm with the same probability for the speakers in all directions. In addition, the speech signal information is different in frequency bands, where the sub-band processing is proposed by the use of the analysis filter bank. The frequency resolution is increased in low frequency components by implementing the proposed filter bank. Then, the affine projection algorithm (APA) is implemented as an adaptive filter on sub-bands that were obtained by the proposed nested microphone array and analysis filter bank. This algorithm adaptively enhances the noisy speech signal. Next, the synthesis filters are implemented for reconstructing the enhanced speech signal. The proposed circular nested microphone array in combination with the sub-band affine projection algorithm (CNMA-SBAPA) is compared with the least mean square (LMS), recursive least square (RLS), traditional APA, distributed multichannel Wiener filter (DB-MWF), and multichannel nonnegative matrix factorization-minimum variance distortionless response (MNMF-MVDR) in terms of the segmental signal-to-noise ratio (SegSNR), perceptual evaluation of speech quality (PESQ), mean opinion score (MOS), short-time objective intelligibility (STOI), and speed of convergence on real and simulated data for white and colored noises. In all scenarios, the proposed method has high accuracy at different levels and noise types by the lower distortion in comparison with other works and, furthermore, the speed of convergence is higher than the compared researches.

Keywords: speech enhancement; adaptive filter; microphone array; sub-band processing; filter bank

1. Introduction

In the current century, the smartphones and other communication devices have been an important part of human life, where it is impossible to have social communications without them [1,2]. One of the principal parts in these smartphones is the signal processing platform. This part has an important role in the telecommunication and audio signal processing. Denoising and dereverberation are two main sections in the signal processing and enhancement platforms, which is the aim of this article, to increase the performance of speech enhancement algorithms [3]. Increasing the number of sensors improves the accuracy of denoising algorithms due to the spatial spectrum extension by providing the proper information. The definition of accuracy in the enhancement algorithms is how the enhanced signal is closer to the original signal with a high level of noise elimination and less distortion. Therefore, the speech enhancement is the main part in such applications as: hearing aid systems, mobile communication, speaker localization and tracking, speech recognition, voice activity detection (VAD), speaker identification, etc. The denoising algorithms should be implemented in a way to keep the speech intelligibility in an acceptable range and to remove a high level of noise and reverberation. Then, the signal-to-noise ratio (SNR) cannot be the only specific factor for comparing the speech enhancement methods. The qualitative criteria such as: perceptual evaluation of speech quality (PESQ) [4], mean opinion score (MOS) [5], and short-time objective intelligibility (STOI) [6] are very useful to show the performance of denoising methods in comparison with other previous works along with quantitative criteria such as: overall SNR and segmental SNR (SegSNR) [7]. The performance of the denoising algorithms is calculated by considering the qualitative and quantitative criteria at the same time, which are the proper measurements for comparison with other previous works.

In recent years, many of the single and multi-channel methods have been proposed for speech enhancement. The single-channel methods are still challenging strategies for the speech enhancement due to the limited information. The traditional speech enhancement methods such as the Wiener filter (WF) and distributed multichannel WF (DB-MWF) [8,9], spectral subtraction [10,11], and statistical-model-based [12,13] have superior performances in stationary noisy environments but the stability and accuracy of these methods are strongly decreased in non-stationary noisy conditions. However, existing noise estimation methods such as minima-controlled recursive averaging [14,15] and minimum statistics [11,16] follow the stationary noise energy. However, they do not have the ability to follow the non-stationary noise energy. For example, the method proposed in [16] is presented to estimate the power spectral density (PSD) of a non-stationary noise signal. This method can be considered in combination with any speech enhancement algorithm, which requires the noise PSD estimation. The presented method follows the spectral minima in each frequency band by minimizing conditional mean square error (MSE) criteria in each time frame, which develops the optimal smoothing parameter for recursive smoothing of the PSD of the noisy speech signal. Therefore, an unbiased noise estimator is presented based on the optimally smoothed PSD estimation and the analysis of the statistics of spectral minima. Therefore, the noise estimation accuracy in some methods [15,16] is affected when the noise is non-stationary. A group of speech enhancement methods are proposed based on a priori information of speech signals such as the auto-regressive hidden Markov model (ARHMM) [17–19]. The noise and speech signals are modeled as an auto-regressive (AR) process in these methods. In addition, the hidden Markov model (HMM) is implemented for modeling the prior information of speech and noise features. For example, the methods in [18,19] are considered for modeling the speech and noise spectrum shape. Therefore, the spectrum gain is calculated instead of the whole spectrum for the speech and noise signals. The noise-spectrum gain estimation is adapted by the fast variations of the signal energy, which is known as non-stationary noise.

Masoud and Sina [20] proposed a novel method based on the normalized fractional of the two-channel least mean square (LMS) algorithm for enhancing the speech signal. The presented algorithm is known as fractional LMS, which is obtained by considering the fractional terms in the calculation of filter coefficients of the standard LMS algorithm. The normalization is a proper strategy to improve the performance of the LMS algorithm. Therefore, a normalization step is implemented on the fractional LMS in order to promote the performance of the enhancement method. The proposed two-channel method has a higher performance in terms of the MSE criteria in comparison with other works. Pagula and Kishore [21] proposed a recursive least square (RLS)-based adaptive filter for the application of speech enhancement. The segmentation step is considered for the microphone signals to provide a better stationary of the speech signals. In the following, the adaptive filter coefficients are calculated based on the modified version of the RLS method. The filter coefficients are calculated in a way to have the least distortion in the enhanced speech signals. The presented method has a high performance in the presence of white noise for a different range of SNRs. Qi et al. [22] proposed a method for estimation of the short-time linear prediction parameters of the Wiener filter. In the presented work, a speech signal spectrum modeling is proposed based on the prior information of the speech linear prediction in order to model the noise as same as the speech signal. The difference between the proposed method with other previous works is the use of multiplicative update rule for better estimation of the coefficients. Tavakoli et al. [23] introduced a framework for the speech enhancement based on an ad-hoc microphone array. A subarray is considered for coherence calculation in the speech signal. A coherence measurement is proposed based on the speech quality in the entrance of the array in order to select the subarrays in the local speech enhancements, when more than one subarray is used. The proposed method is evaluated based on quantitative and qualitative criteria such as: array gains, speech distortion ratio, PESQ, and STOI to show the superiority of the algorithm. Shimada et al. [24] proposed an unsupervised speech enhancement method based on the non-negative matrix factorization and sub-band beamforming for robust speech recognition against the noise. In the recent years, the minimum variance distortionless response (MVDR) beamforming is widely used to achieve the speech enhancement because this method properly works when there are steering vectors for the speech signal and spatial covariance matrix for the noise. In the presented algorithm, an unsupervised method decomposes each time-frequency bin to the sum of the noise and signal by implementing the multi-channel non-negative matrix factorization (MNMF). The presented method estimates the spatial covariance matrix (SCM) for the signal and noise by the use of spectral noise and speech features. In this paper, the online MVDR beamforming is proposed via an adaptive update for the MNMF parameters. Kavalekalam et al. [25] proposed a speech enhancement model-based method to increase the speech perception for auditory earphones applications. In the proposed method, a binaural speech enhancement framework is introduced, which is implemented by a speech production approach. The proposed speech enhancement framework is based on a Kalman filter, which is presented to use the speech production dynamic in the procedure of the speech enhancement. The Kalman filter needs to have an estimation from the short time predictor (STP) of clean speech, noise, and the pitch estimation of the clean speech. A binaural method for STP parameters estimation is proposed in this paper with a directional pitch predictor based on the harmonic model and maximum likelihood (ML) criteria for pitch features estimations. These parameters are calculated just based on 2-microphones signals equivalent to human ears. Botinhao et al. [26] proposed a simultaneous noise-reverberation enhancement method for text-to-speech (TTS) systems. The recorded voices in noisy-reverberant environments affects the quality of the TTS systems. A simple way is to increase the quality of the prerecorded speech signals for the TTS training system by speech enhancement methods such as: noise suppression and dereverberation algorithms. Then, a recurrent neural network is considered in this paper for the speech enhancement. The neural network is trained by parallel data of clean speech and recorded speech with low quality. The low quality speech signal is obtained by the addition of environmental noise and convolution between the room impulse response and the clean speech. The separated neural networks are trained by only-noise, only-reverberation, and noisy-reverberant data. The quality of the training data with a low quality speech signal is highly improved by the use of this neural network. Wang et al. [27] proposed a model-based method for speech enhancement in modulation domain by the use of a Kalman filter. The proposed predictor models the estimated amplitude spectral dynamically from the speech and noise to calculate the minimum mean square error (MMSE) of the speech amplitude spectrum taking into account that the noise and speech are additive in the complex plane. The stationary Gaussian model is proposed to consider the dynamic noise amplitude as same as the dynamic speech amplitude, which is a mixture of Gaussian models that the centers are located in a complex plane.

In our article, a multi-channel speech enhancement method is introduced based on the proposed circular nested microphone array in combination with the sub-band affine projection algorithm (CNMA-SBAPA). A nested microphone array increases the accuracy of the speech enhancement methods by increasing the information. Nevertheless, spatial aliasing is one of the challenges when microphone arrays are used. Firstly, a uniform circular nested microphone array (CNMA) is proposed for eliminating the spatial aliasing. Additionally, the array dimensions are designed in a way to be applicable in the real conditions. The speech components are variable in frequency bands. Therefore, a sub-band processing method is considered for speech signals. This method provides the high frequency resolution in low speech frequency components. Finally, the affine projection algorithm (APA), as an adaptive method for the speech enhancement, is implemented on sub-band signals from the circular nested microphone array (NMA). Since each APA block is implemented on a sub-band with specific information, the accuracy and speed of convergence are increased in this condition. In the last step, the synthesis filters are used to generate the enhanced speech signal. The proposed system with sub-band APA is compared by the quantitative (segmental SNR), qualitative (PESQ, MOS, and STOI) criteria, and speed of convergence with the least mean square (LMS), traditional APA, recursive least square (RLS), distributed multichannel Wiener filter (DB-MWF), and multichannel nonnegative matrix factorization-minimum variance distortionless response (MNMF-MVDR) algorithms on real and simulated data under white and colored noisy conditions. The results show the superiority of the proposed system in comparison with other previous works in all environmental conditions.

Section 2 shows the microphone signal model and the proposed uniform circular nested microphone array. Section 3 includes the proposed sub-band algorithm with analysis and synthesis filter banks in combination with the sub-band APA. The results on real and simulated data are discussed in Section 4. Section 5 includes some conclusions.

2. The Microphone Model and Proposed Nested Microphone Array

In this section, the microphone signal model was presented to produce the simulated data. In addition, the uniform CNMA was proposed for eliminating the spatial aliasing. Additionally, the nested subarrays and microphone combinations are introduced in this section.

2.1. Microphone Signal Model

The microphone signal modeling is an important part in the implementation of speech processing algorithms such as: speech enhancement, speaker tracking, speech recognition, etc. Two models are usually considered in this processing: ideal and real models [28]. In the ideal model, which is known as an open-space model, the received signal in a microphone place is a weakened and delayed version of the transmitted signal from the source location. The ideal model for microphone signals is expressed as:

$$x_m[n] = \frac{1}{r_m}s[n - \tau_m] + v_m[n], \qquad (1)$$

where $x_m[n]$ is the received signal in the m-th microphone, $s[n]$ is the speech source signal (transmitted signal), r_m is the distance between source and m-th microphone, τ_m is the time delay between source and

m-th microphone, and $v_m[n]$ is the additive noise in m-th microphone place. This model cannot show the real environments and close space conditions because the reverberation effect is discarded. Therefore, the real model is introduced for microphone signal simulations to provide the real environmental conditions for evaluating the speech enhancement algorithms. The real model simulates the microphone signal similar to the environmental conditions. The expression for real model is shown as:

$$x_m[n] = s[n] * \gamma_m[r_m, n] + v_m[n], \quad (2)$$

where the source signal is convolved to the room impulse response to model the real environments. In this equation, $\gamma_m[r_m, n]$ is the impulse response between the source and m-th microphone, which contains the attenuation factor and whole reverberation effect in the real conditions, and $*$ denotes to the convolution operator. The simulated signals are similar to real conditions by considering this mathematical real model.

2.2. The Proposed Uniform Circular Nested Microphone Array

The microphone array increases the accuracy of the speech enhancement algorithms due to increasing the information. However, the spatial aliasing based on the inter-microphone distances destroys the recorded speech signals, and in the following, the performance of the speech enhancement algorithms. Nested microphone array has the capability to eliminate the spatial aliasing [29]. In this section, a uniform CNMA is proposed where by having a symmetrical shape, provides the same probability for all speakers around the array, and the quality of the enhanced signals are not dependent on the position of the speakers. Additionally, its small structure helps to be applicable in most of the conditions in comparison with other big arrays. Figure 1 shows the block diagram of the proposed speech enhancement algorithm, where the NMA part with its analysis filters and down-sampler blocks are shown in the left side.

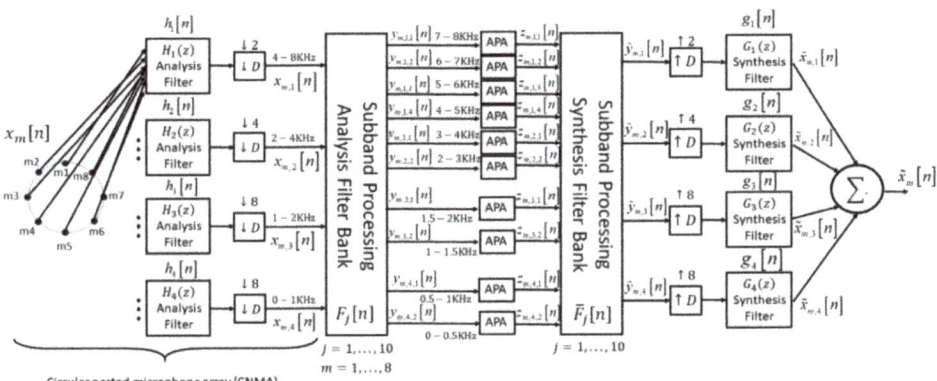

Figure 1. The block diagram of the proposed circular nested microphone array in combination with the sub-band affine projection algorithm (CNMA-SBAPA) for the speech enhancement.

The speech signal has a frequency range of [0–8000] Hz with a sampling frequency. $F_s = 16,000$ Hz The proposed CNMA is designed for the frequency range [50–7800] Hz, which covers the wideband speech spectrum. The CNMA is structured by four subarrays. The first subarray is designed for the range B1 = [3900–7800] Hz, of central frequency $f_{c1} = 5850$ Hz. The inter-microphone distance ($d_{\lim}$) should be $d_{\lim} < \lambda/2$ (λ is the wavelength of the highest frequency component in the related sub-band) to avoid the spatial aliasing, this is $d_{\lim(1)} < 2.2$ cm for the first subarray. The second subarray covers the frequency range B2 = [1950–3900] Hz with a central frequency of $f_{c2} = 2925$ Hz, therefore $d_{\lim(2)} = 2d_1 < 4.4$ cm. The third subarray is defined for the frequency range B3 = [975–1950] Hz with a central frequency of $f_{c3} = 1462$ Hz and $d_{\lim(3)} = 4d_1 < 8.8$ cm. Finally, the forth subarray is designed for the frequency range B4 = [50–975] Hz with a central frequency of $f_{c4} = 512$ Hz and the inter-microphone distance is $d_{\lim(4)} = 8d_1 < 17.6$ cm. For a more complexity system, a higher number of microphones could be considered to design a larger nested microphone array. Table 1 shows the summarized information to design the uniform CNMA.

Table 1. The information to design the proposed uniform CNMA.

Band	Bandwidth	Analysis Filter Bank	f_c	$d_{\lim}$
1	B1 = [3900–7800] Hz	B1,1 = [6825–7800] Hz B1,2 = [5850–6825] Hz B1,3 = [4875–5850] Hz B1,4 = [3900–4875] Hz	5850 Hz	<2.2 cm
2	B2 = [1950–3900] Hz	B2,1 = [2925–3900] Hz B2,2 = [1950–2925] Hz	2925 Hz	<4.4 cm
3	B3 = [975–1950] Hz	B3,1 = [1425–1950] Hz B3,2 = [975–1425] Hz	1462 Hz	<8.8 cm
4	B4 = [50–975] Hz	B4,1 = [512–975] Hz B4,2 = [50–512] Hz	512 Hz	<17.6 cm

The microphone array was structured to have the closest microphone distances as $d_{sim(1)} = 2.2$ cm (for the simulated data) based on the designed CNMA. Therefore, the first subarray included the microphone pairs {1,2}, {2,3}, {3,4}, {4,5}, {5,6}, {6,7}, {7,8}, and {8,1}. The microphone pairs {1,3}, {3,5}, {5,7}, {7,1}, {2,4}, {4,6}, {6,8}, and {8,1} were selected for the second subarray with an inter-microphone distance of $d_{sim(2)} = 4.2$ cm. The third subarray has the inter-microphone distance of $d_{sim(3)} = 5.6$ cm. Then, the microphone pairs {1,4}, {2,5}, {3,6}, {4,7}, {5,8}, {6,1}, {7,2}, and {8,3} wereconsidered for this subarray. For the last subarray, the inter-microphone distance is $d_{sim(4)} = 6$ cm and the microphone pairs {1,5}, {2,6}, {3,7}, and {4,8} were selected for the implementation. Given our actual microphone array, the minimum inter-microphone distance that we could have was2.7cm (for the real data). For this reason, we did two evaluations, one for simulated data with $d_{sim(1)} = 2.2$ cm, as dictated by the theory, and one for real data with $d_{real(1)} = 2.7$ cm, to match our hardware. All subarrays are shown in Figure 2, which shows the designed CNMA with its small shape.

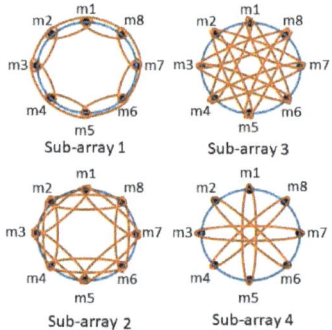

Figure 2. The proposed uniform CNMA and allocated microphones for each subarray.

Each subarray needs an analysis filter bank to avoid the spatial aliasing and imaging. Figure 1 (left and right sides) shows the analysis and synthesis filter banks along with the up-sampler and down-sampler blocks. The multirate sampling by the use of up-samplers and down-samplers is implemented to provide the frequency bands. As shown in Figure 3a, the analysis filter bank $H_i(z)$ and down-sampler D_i are realized as a multi-level tree structure. Each stage of the tree requires a high-pass filter (HPF) $HP_i(z)$, a low-pass filter (LPF) $LP_i(z)$, and a down-sampler D_i (for the analysis filter bank) or up-sampler D_i (for the synthesis filter bank). The relation between the analysis filter bank $H_i(z)$, the LPFs, and HPFs in the tree structure is expressed as:

$$\begin{aligned} H_1(z) &= HP_1(z) \\ H_2(z) &= LP_1(z)HP_2(z^2) \\ H_3(z) &= LP_1(z)LP_2(z^2)HP_3(z^4) \\ H_4(z) &= LP_1(z)LP_2(z^2)LP_3(z^4). \end{aligned} \tag{3}$$

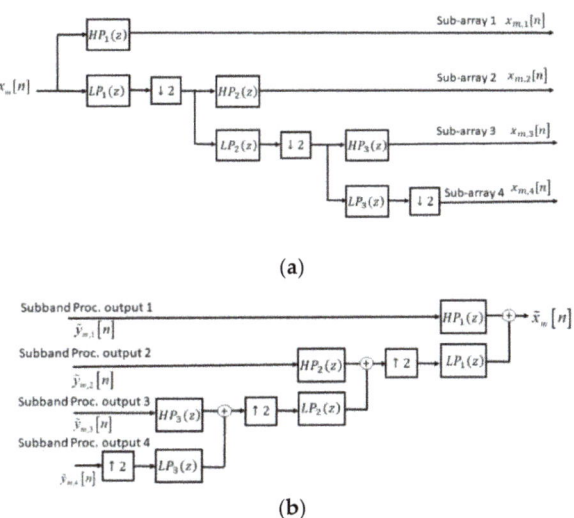

(a)

(b)

Figure 3. The tree structure for (**a**) analysis and (**b**) synthesis filters in CNMA.

The synthesis filters $G_i(z)$ are the mirror image of analysis filters $H_i(z)$, which are implemented by the tree structure as seen in Figure 3b.

In each level of the tree, a 52-tap finite impulse response (FIR) LPF and HPF are implemented by the Remez method. The parallel filters have a stop-band attenuation of 50dB and a transition band 0.0575. Figure 4 shows the frequency response for the analysis filter banks.

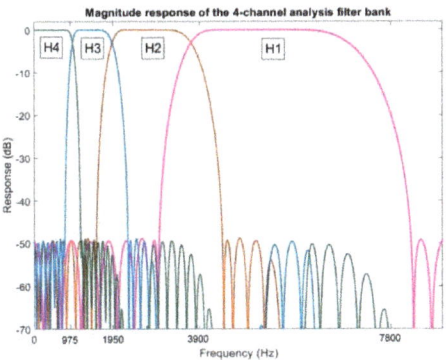

Figure 4. The frequency response for the analysis filter banks.

3. The Proposed Multiresolution Sub-band-APA for the Speech Enhancement

Speech is a wideband and non-stationary signal, where each frequency band has different information. This feature for the speech signal provides the conditions to evaluate the speech spectrum components by considering different frequency resolution. For example, speech information is condensed at the lower part of the spectrum. Therefore, the accuracy of the speech enhancement algorithm is increased by a focus to low frequency components. In this article, a specific sub-band processing along with a filter bank was proposed for paying more attention to lower frequencies by the use of filters with narrower bandwidths. Table 2 shows the information to design and implement this analysis filter bank. There is still not any certain rule for selecting the number of frequency bands. Of course, by having narrower band filters in low frequencies, we have more frequency resolution, but the concern is the computational complexity. In other hands, adding each more filter means entering more microphone pairs and more calculations. Based on the experiments, this number of frequency bands prepares enough performance and acceptable level of complexity.

Table 2. The required information to design the analysis filter bank for sub-band processing in the proposed CNMA-SBAPA algorithm.

Filters	Bandwidth (Hz)	f_{min} (Hz)	f_{max} (Hz)	Filter Length (Samples)
$F_1[n]$	462	50	512	93
$F_2[n]$	462	512	975	115
$F_3[n]$	450	975	1425	102
$F_4[n]$	525	1425	1950	124
$F_5[n]$	975	1950	2925	109
$F_6[n]$	975	2925	3900	118
$F_7[n]$	975	3900	4875	131
$F_8[n]$	975	4875	5850	140
$F_9[n]$	975	5850	6825	146
$F_{10}[n]$	975	6825	7800	151

As seen, the filter bandwidth is smaller in low frequencies in comparison with high frequencies. This property increases the frequency resolution for low frequencies. The most important benefit in sub-band processing is the noise estimation from the silent part of the speech signal in each sub-band. Since in the proposed denoising method, the noise estimation is required as an input for the enhancement algorithm. Therefore, the more accurate and stationary noise estimation is obtained by sub-band processing of the speech signal, which increases the denoising algorithm performance. If $x_m[n]$ is considered as an input signal for the m-th microphone, the analysis filter output for the CNMA is expressed as:

$$x_{m,i}[n] = x_m[n] * h_i[n] \text{ where } \{m = 1,\ldots,8 \text{ and } i = 1,\ldots,4\}, \quad (4)$$

where $x_{m,i}[n]$ is the analysis filter output and $h_i[n]$ is the impulse response for this filter. Therefore, the spatial aliasing is eliminated from each microphone pairs of CNMA by the use of analysis filters, which are designed specifically for each subarray. In the following, the microphone signals are entered to the proposed analysis filter bank for the sub-band processing. As shown in Table 2, each microphone signal is divided into 10 sub-bands. These numbers of sub-bands were selected based on our experiments in order to provide a proper efficiency and with low computational complexity, by preparing a high frequency resolution in low frequencies. Therefore, the output of the proposed analysis filter bank is expressed as:

$$y_{m,i,j}[n] = x_{m,i}[n] * F_j[n] \text{ where } \{j = 1,\ldots,10, i = 1,\ldots,4, m = 1,\ldots,8\}, \quad (5)$$

where $F_j[n]$ is the impulse response for each sub-band filter in the analysis filter bank and $y_{m,i,j}[n]$ is the output of the analysis filter bank for the j-th sub-bands and m-th microphone. The signals

$y_{m,i,j}[n]$ are the sub-band microphone signals for the proposed sub-band-APA algorithm. In the following, the sub-band-APA (SBAPA) algorithm along with the circular nested microphone array (CNMA-SBAPA) is proposed for the speech enhancement. Adaptive filters as an important tool in digital signal processing have been utilized for many years in such application as: speech signal enhancement, system identification, localization and tracking, etc. In adaptive filters, the coefficients change periodically to be adapted based on the time varying features of the noise, and this property increases the performance of the denoising system in comparison with normal methods. In addition, these filters are non-linear and homogeneous since their features are dependent on the input signal. The adaptive filters have the following advantages: low delay and better tracking in non-stationary conditions [30]. These advantages are very important in dereverberation, denoising, time delay estimation, channel equalization, and speaker tracking applications. In these applications, low delay and robustness against of non-stationary noisy and reverberant conditions are important parameters to improve the performance of the proposed systems. The existence of the reference signal, which is hidden in the filter coefficient estimations, defines the system performance. Figure 5 shows the general structure of the adaptive filter in denoising applications.

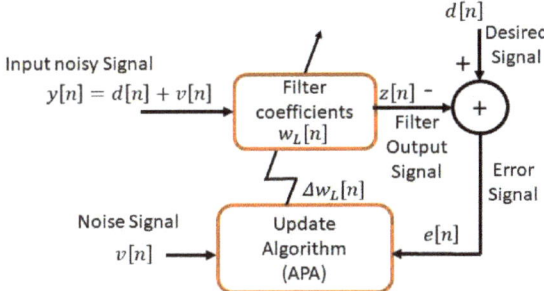

Figure 5. The general structure of the adaptive filter for denoising applications.

We change the notation for input signal in adaptive filter ($y_{m,i,j}[n]$) to $y[n]$ for simplifying the mathematical expressions. An adaptive filter is expressed as follows [31]:

$$z[n] = w_L[n] * y[n], \tag{6}$$

where n is the time index, $z[n]$ is the adaptive filter output, and $w_L[n]$ is the adaptive filter coefficients with length L. The update algorithm in Figure 5 is considered as a principal part for an adaptive filter, which is the APA in this article. The main idea for an adaptive filter is to minimize the error signal $e[n]$ to make the output of the filter as similar as the desired signal.

The input signal $y[n]$ for the adaptive filter is considered as the summation of the noise ($v[n]$) and desired signal ($d[n]$), which is described as:

$$y[n] = d[n] + v[n]. \tag{7}$$

The adaptive filter has a FIR structure, namely the filter is designed based on the limited number of coefficients in the time domain. For a filter with order of L, the filter coefficients are defined as:

$$w_L[n] = [w[0], w[1], \ldots, w[L-1]]. \tag{8}$$

The error signal or cost function is defined as the difference between estimated and desired signal, namely:

$$e[n] = d[n] - z[n]. \tag{9}$$

As shown in Equation (6), the output of the adaptive filter $z[n]$ is defined as the convolution between the filter coefficients $w_L[n]$ and the input signal $y[n]$, where $y[n]$ is considered as the input of the adaptive filter, namely:

$$y[n] = [y[n], y[n-1], \ldots, y[n-L]]. \tag{10}$$

In addition, the adaptive filter coefficients change during the time, which is written as:

$$w_L[n] = w_L[n-1] + \Delta w_L[n], \tag{11}$$

where $\Delta w_L[n]$ is defined as the correction factor for the filter coefficients. The adaptive filter produces the correction factor based on the input and error signal. In Figure 5, several algorithms can be considered for updating the filter coefficients. The APA is one of the fastest and most efficient methods for this purpose. The AP algorithms were introduced to improve the speed of convergence in the gradient-based algorithms, especially when the input signal has a non-stationary spectrum. It is because the speed of convergence is decreased in the case of non-stationary and constraint spectrums [30].

Filter update equation is one of the most important features in the AP algorithms, which uses N vectors of the input data to update the filter coefficients instead of using one vector of the input data, i.e., the normalized least mean square (NLMS). Therefore, more information was considered in the time for accurately updating the filter coefficients. Thus, the AP algorithm is known as an improved and extended version of the NLMS method or it can be expressed mathematically as a constraints minimization problem, which is expressed as follows.

The variation for L filter coefficients during the two consecutive times is given by:

$$\Delta w_L[n] = w_L[n] - w_L[n-1]. \tag{12}$$

We minimized Equation (13) under N constraints, which are shown in Equation (14) to extend the adaptive filter algorithm.

$$\|\Delta w_L[n]\|^2 = \Delta w_L^T[n] \Delta w_L[n], \tag{13}$$

where N constraints are defined as follows:

$$w_L^T[n] y[n-k] = d[n-k] \text{ for } k = 0, \ldots, N-1, \tag{14}$$

where $y[n-k]$ is the vector of N last sample from the input signal and $d[n]$ is the desired signal, see Figure 5. The proposed solution formulates the update algorithm for AP, which is expressed as:

$$w_L[n] = w_L[n-1] + A^T[n] \left(A[n] A^T[n] \right)^{-1} e_N[n], \tag{15}$$

where:

$$A[n] = (y_L[n], y_L[n-1], y_L[n-2], \ldots, y_L[n-N+1])^T, \tag{16}$$

and $e_N[n]$ is a vector of size $N \times 1$, which is written as:

$$e_N[n] = d_N[n] - A[n] w_L[n-1]. \tag{17}$$

The vector $d_N[n]$ is the desired signal with size $N \times 1$, namely:

$$d_N[n] = (d[n], d[n-1], \ldots, d[n-N+1])^T. \tag{18}$$

The general format for AP algorithm is obtained by rewriting Equation (15) as:

$$w_L[n] = w_L[n-1-\alpha(N-1)] + \mu A_\tau^T[n] \left(A_\tau[n] A_\tau^T[n] + \delta I \right)^{-1} e_{N\tau}[n]. \tag{19}$$

If $e_{N\tau}[n]$ is considered as $e_{N\tau}[n] = d_{N\tau}[n] - A_\tau w_L[n-1-\alpha(N-1)]$, then:

$$A_\tau[n] = (y_L[n], y_L[n-\tau], \ldots, y_L[n-(N-1)\tau])^T, \quad (20)$$

and the signal $d_{N\tau}[n]$ is expressed as:

$$d_{N\tau}^T[n] = (d[n], d[n-\tau], \ldots, d[n-(N-1)\tau]). \quad (21)$$

As shown in Equation (19), the N required vectors to update the adaptive filter are not necessarily to be the last data vectors. Therefore, several versions of AP algorithms are defined based on the way to select the input data and parameters in Equation (19). There are some developed algorithms based on these parameters selections such as: the NLMS along with the orthogonal correction factor (OCF-NLMS) [32], the partial rank affine projection algorithm (PRAPA) [33], and the standard APA [34] whose parameters are $\alpha = 0$, $\delta = 0$, $\tau = 1$. If δ parameter differs to 0, the APA algorithm is extended to APA with regularization (R-APA) [35], where the update equation for the filter coefficients is a specific case of the Levenberg Marquardt regularized APA (LMR-APA) algorithm [36].

The introduced AP algorithm contains one input signal. Since pairs of microphones are used in the proposed CNMA, the AP algorithm is generalized to a two-microphone version [37]. Firstly, the generalization of a two-microphone structure is defined, where each microphone contains the mixing speech and noise signal, which is expressed as (see Figure 6a):

$$q_m[n] = \sum_{i=1}^{2} \sum_{r=1}^{L-1} \mathbf{p}_{im}[r] s_i[n-r], \, m = 1, 2, \quad (22)$$

where $s_i[n]$ represents the source signals, $q_m[n]$ is the microphone signals, L is the impulse response length, and $\mathbf{p}_{im}[r]$ are the impulse responses between the microphone and sources. These impulse responses are considered as linear time-invariant (LTI) systems. Two source signals $s_i[n]$ are selected as the speech signal $s[n]$ and noise signal $b[n]$. It is assumed that the speech and noise signals are independent, which means $E\{s[n]b[n-m]\} = 0$, $\forall m$, where E denotes to expected value. Then, the noise and speech signals are uncorrelated. Based on the general structure, which is shown in Figure 6a, the microphone signals $q_1[n]$ and $q_2[n]$ are expressed as follows:

$$q_1[n] = s[n] * \mathbf{p}_{11} + b[n] * \mathbf{p}_{21}, \quad (23)$$

$$q_2[n] = s[n] * \mathbf{p}_{12} + b[n] * \mathbf{p}_{22}. \quad (24)$$

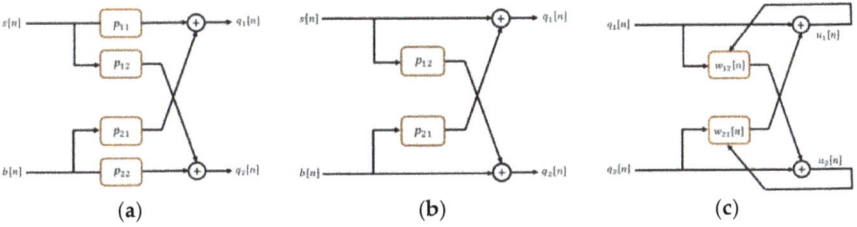

Figure 6. (a) The general structure of the proposed denoising system, (b) the simplified presented model for the two-microphone system, and (c) the affine projection algorithm (APA) structure for the two-microphone.

In addition, $\mathbf{p}_{11}$ and $\mathbf{p}_{22}$ represent the impulse responses for direct path, and $\mathbf{p}_{12}$ and $\mathbf{p}_{21}$ are cross-coupling for the channels between the sources and microphones. The presented model is simplified by considering $\mathbf{p}_{11} = \mathbf{p}_{22} = \delta[n]$, which is shown in Figure 6b as:

$$q_1[n] = s[n] + b[n] * \mathbf{p}_{21}, \tag{25}$$

$$q_2[n] = s[n] * \mathbf{p}_{12} + b[n]. \tag{26}$$

Therefore, the microphone signals are generated based on the impulse responses between the source and microphones, noise, and speech signals. The structure in Figure 6c was proposed to retrieve the source signal from the received noisy signals $q_1[n]$ and $q_2[n]$. The proposed structure provides the conditions to retrieve the original signal by the use of adaptive filters $\mathbf{p}_{11}$ and $\mathbf{p}_{22}$. The signals $u_1[n]$ and $u_2[n]$ for the two-microphone structure are defined as follows:

$$u_1[n] = q_1[n] - q_2[n] * \mathbf{w}_{21}[n], \tag{27}$$

$$u_2[n] = q_2[n] - q_1[n] * \mathbf{w}_{12}[n], \tag{28}$$

where in Equations (27) and (28), $\mathbf{w}_{12}[n]$ and $\mathbf{w}_{21}[n]$ are the adaptive filters for eliminating the noise of microphone signal $q_1[n]$ and the speech of microphone signal $q_2[n]$, respectively. Signals $u_1[n]$ and $u_2[n]$ are rewritten by replacing Equations (25) and (26) to Equations (27) and (28) as:

$$u_1[n] = s[n] * [\delta[n] - \mathbf{p}_{12} * \mathbf{p}_{21}] + b[n] * [\mathbf{p}_{21} - \mathbf{w}_{21}[n]], \tag{29}$$

$$u_2[n] = b[n] * [\delta[n] - \mathbf{p}_{21} * \mathbf{p}_{12}] + s[n] * [\mathbf{p}_{12} - \mathbf{w}_{12}[n]]. \tag{30}$$

Two adaptive filters $\mathbf{w}_{12}[n]$ and $\mathbf{w}_{21}[n]$ are required to retrieve the original speech signal from the noisy signals $u_1[n]$ and $u_2[n]$. There is just a unique structure for adaptive filters $\mathbf{w}_{12}[n]$ and $\mathbf{w}_{21}[n]$ as $\mathbf{w}_{12}[n] = \mathbf{p}_{12}$ and $\mathbf{w}_{21}[n] = \mathbf{p}_{21}$ to retrieve the enhanced speech of noisy signals $u_1[n]$ and $u_2[n]$. This structure requires a VAD for preparing the noise estimation from the silent part of the recorded signals.

The AP algorithm is generalized to a two-microphone structure based on the obtained Equation (19) for updating the filter coefficients. The AP algorithm is the generalized version of the two-microphone NLMS [38], which is shown in Figure 6c for adaptive speech enhancement algorithm. Therefore, the adaptive filter coefficients $\mathbf{w}_{12}[n]$ and $\mathbf{w}_{21}[n]$ for two-microphone APA are expressed as:

$$\mathbf{w}_{12}[n] = \mathbf{w}_{12}[n-1] + \frac{\mu_{12}}{\mathbf{q}_1[n]\mathbf{q}_1[n]^T + \delta I} \mathbf{q}_1[n] u_2[n], \tag{31}$$

$$\mathbf{w}_{21}[n] = \mathbf{w}_{21}[n-1] + \frac{\mu_{21}}{\mathbf{q}_2[n]\mathbf{q}_2[n]^T + \delta I} \mathbf{q}_2[n] u_1[n], \tag{32}$$

where $\mathbf{q}_1[n]$ and $\mathbf{q}_2[n]$ are defined as $\mathbf{q}_1[n] = [q_1[n], q_1[n-1], \ldots, q_1[n-N+1]]$ and $\mathbf{q}_2[n] = [q_2[n], q_2[n-1], \ldots, q_2[n-N+1]]$. The matrices of the two-microphone signals $q_1[n]$ and $q_2[n]$ have dimensions $L \times N$, where L is the adaptive filter length and N is the projection order. The two parameters μ_{12} and μ_{21} are the step sizes, which control the convergence of adaptive filters $\mathbf{w}_{12}[n]$ and $\mathbf{w}_{21}[n]$. These parameters should be selected in the range [0,2] to assure the convergence of AP algorithm. If N is selected as 1, the AP algorithm is converted to the NLMS method.

The proposed sub-band APA not only increased the accuracy of the speech enhancement algorithm, but also the speed of convergence was improved (Table 6 in the results section) in the implementations because the noise was estimated separately for each sub-band and it was stationary on narrow bandwidths. Then, the SBAPA was implemented on generated sub-bands by the analysis filters in Figure 3. As shown in Figure 1, a symmetrical synthesis filter bank and synthesis filters related to the nested microphone array were implemented for the reconstruction the final enhanced signal.

The synthesis filters as similar as the analysis filters were implemented based on the tree structure in Figure 3b. Finally, all sub-band signals were summed to generate the final enhanced signal. In the next section, the performance of the proposed CNMA-SBAPA was compared with other previous works.

4. Results and Discussion

The experiments in order to evaluate the performance of the proposed method were implemented on the real and simulated data. The TIMIT dataset was considered for the simulated data, where the data collection MDAB0 by four continuous sentences SX139, SX229, SX319, and SX409 were selected as a male speaker in the simulations [39]. This dataset includes short sentences for testing and training the algorithms. The tones and frequency components are two different parameters in the speech signal. There are pitch and speech spectrum components for the speakers. It is important to work with male or female signals for the algorithms, which works with the pitch parameter. Since this parameter changes highly based on the gender. Since we consider the speech spectrum, then the issue to use the male or female speakers does not change the results. Therefore, 12.5 s male-speech signal is used for implementations and experiments. A voice activity detector is implemented to detect the silence part of the speech signal [40], and the noise spectrum is estimated of these parts for the proposed SBAPA. Figure 7 shows the simulated room with the location of speakers and microphone array. The inter-microphone distances $d_{sim} = 2.2$ cm for the simulated data was selected based on the designed array. A speaker and a steered noise source were considered in the simulations. The room dimensions, speaker, and noise source locations were selected as 475,592,420cm, 374,146,110cm, and 362,412,120cm, respectively. These dimensions and locations were considered the same as the real room recording conditions. In addition, the proposed algorithm was implemented on real data to evaluate the real effect of the noise and reverberation on the performance. For this purpose, the real speech signal was recorded in the speech processing laboratory at Fondazione Bruno Kessler (FBK), Trento, Italy. Figure 8 shows a view of the recording room at FBK. Two electronic speakers were used instead of the human and noise source in the process of data recording. In addition, Figure 8 shows the position of the circular NMA in the center of the room. We were able to consider the minimum inter-microphone distance in the real conditions with our setup (see Figure 8) as $d_{real} = 2.7$ cm because of the microphone dimension, electronic board, and the microphone shield. Additionally, each microphone had a cross section, where in the real conditions it was about 0.7cm. It means it is hard to measure the exact distance between two microphones and it has some errors. Since all cross sections in a microphone are areas for a sound recording, then, based on all limitations, we were forced to have this inter-microphone distance for real data implementation even with a few millimeters difference with the mathematical calculations. Therefore, the differences in the results of our proposed method for the real and simulated data were for this an issue. In the real condition, there are always some inaccuracy factors for the measurements. We found the center of the room and the microphones were located on the table based on the primary measurements. All microphones were connected to the sound recording system, which uses parallel acquisition for all microphone channels. All channel acquisitions were synchronized and there was not any delay between recorded signals in different microphones or channels. The phase error based on the recording condition was very low and was even close to zero based on the audio recording system. In the real room, the table did not make any direct reflection. All the reflected waves from the table will cross to the walls and ceiling firstly, and since all of them were covered with curtains and sound absorption panels, the indirect reflections to the microphones were very few. Both speakers were connected to the two computers for playing the speech and noise with a sampling frequency of $F_s = 16,000$ Hz. The microphone, sound, and noise sources were selected in the simulations with exactly the same real conditions for the results to be comparable in these two conditions. Figure 9 shows the time-domain and spectrum of the male speech signal.

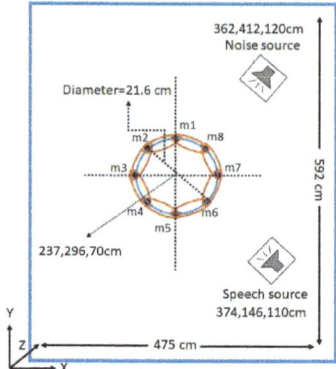

Figure 7. A view of the simulated room with the positions of speakers, noise source, and microphones.

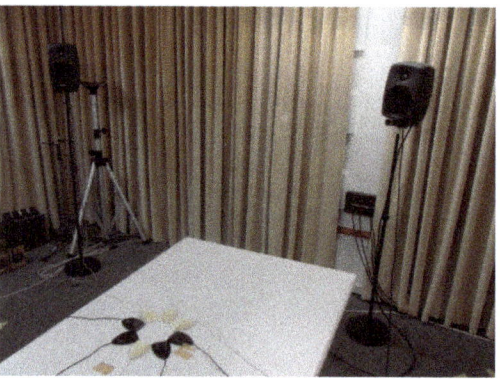

Figure 8. The real recording room in the speech processing laboratory, FBK, Trento, Italy.

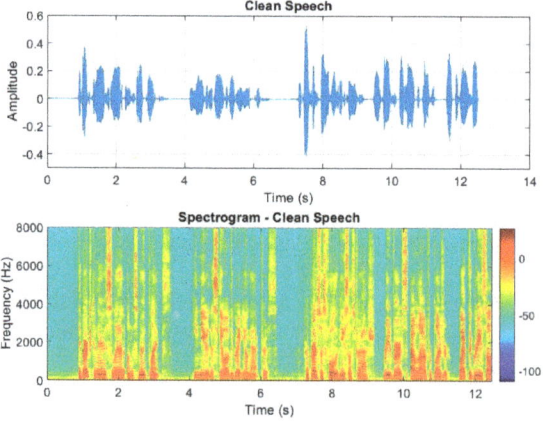

Figure 9. The time-domain and spectrum for the male speech signal.

The reverberation effect was considered in the experiments to provide the simulation conditions similar to the real scenarios. The image model was implemented in the simulations to produce the reverberation effect similar to the real conditions [41]. The image model produced the room impulse response between the source and microphone by considering the speaker position, microphone location, sampling frequency, room dimension, room reflection coefficients, impulse response length, and reverberation time. The received signal to the microphone was simulated by the convolution between the generated impulse response by the image method and the source signal. The impulse response was generated for the noise and speech sources because both receive the same effect of the room reverberation. In addition, noise was additive with the speech signal in the microphone positions. The room reverberation time was selected as $RT_{60} = 350$ ms, which was considered for a room with a low level of reverberation to be the same as the real conditions. To generate the noisy signal, five types of noise were considered for the simulated and real data such as white noise, babble noise, train noise, car noise, and restaurant noise. Figure 10 shows the time-domain and spectrum for these noisy signals according to a SNR=0dB. The noise signal duration was 12.5 s, the same as the speech signal.

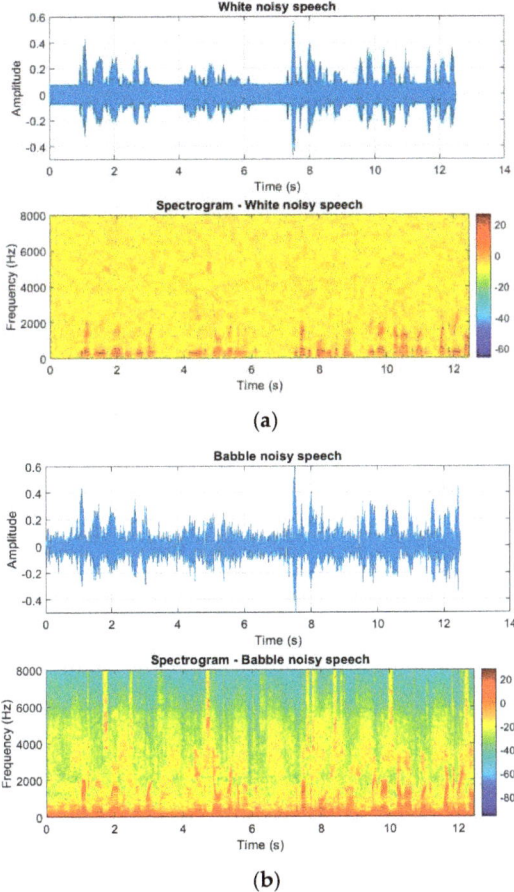

(a)

(b)

Figure 10. *Cont.*

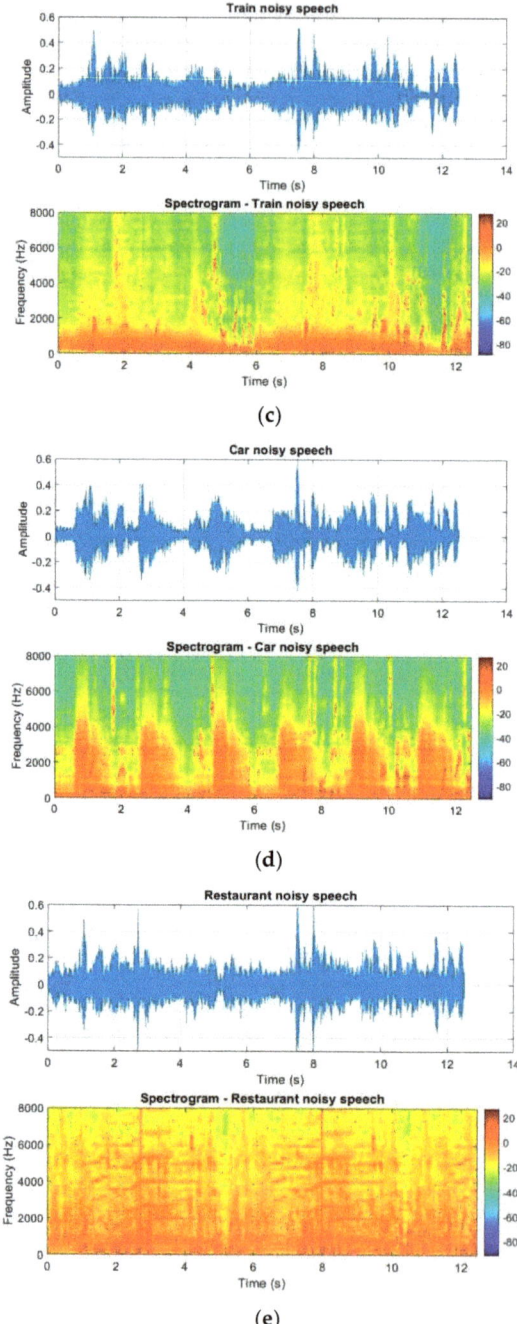

Figure 10. The time-domain and spectrum for a noisy speech signal with (**a**) white noise, (**b**) babble noise, (**c**) train noise, (**d**) car noise, and (**e**) restaurant noise according to a signal-to-noise ratio (SNR) = 0 dB.

The Hamming window with a length of 30 ms was selected for signal blocking to keep the stationarity of the signal in the short time. The projection order was considered as $N = 4$ to keep the computational complexity in an acceptable range in addition to a proper accuracy of the algorithm. Additionally, the step sizes were chosen as $\mu_{12} = 1$ and $\mu_{21} = 1$ to provide the fast convergence for the proposed SBAPA in the real-time implementations. The evaluations in this article were implemented by the use of MATLAB software version 2019b on a PC with processor Inter Core i7-7700k, 4.20 GHz, and with 32GB RAM to be able to implement the proposed algorithm in the real-time conditions.

The proposed SBAPA in combination with a proposed circular nested microphone array (CNMA-SBAPA) was compared with the LMS [20], traditional APA [31], RLS [21], DB-MWF [9], and MNMF-MVDR [24] algorithms. These methods were compared because all of them are based on the adaptive filters and multi-channel beamforming as a main category for comparison. There are many methods for comparison with the proposed algorithm but the comparison should be based on the common theme in implementations. Therefore, the adaptive filter-based algorithms were selected for this comparison. The qualitative and quantitative criteria were considered to show the superiority of the proposed method in comparison with other previous works. For this purpose, the SegSNR [7], PESQ [4], MOS [5], and STOI [6] criteria were selected for the comparison. The SegSNR is a quantitative criterion, which shows the improvement in the enhanced signal due to the percentage of the noise power elimination from the noisy signal, namely:

$$\text{SegSNR}_{(dB)} = \frac{1}{R} \sum_{i=0}^{R} 10 \log_{10} \left(\frac{\sum_{j=0}^{Q-1} |S_j[n]|^2}{\sum_{j=0}^{Q-1} |S_j[n] - Z_j[n]|^2} \right) VAD_i \tag{33}$$

where $S[n]$ and $Z[n]$ are the clean and enhanced speech signals, respectively. The variable Q is the mean averaging value of the SNR for the output signal. The variable R is the number of only-speech frames and VAD is a speech detector, which is 1 for only-speech frames and 0 for only-noise frames. Therefore, the SegSNR is appropriate to show the speech enhancement performance. Many of the speech enhancement algorithms eliminate some part of the speech signals in addition to the noise frames, which decreases the speech perception for the enhanced signals. Then, three well-known qualitative criteria are considered in the evaluations. The first one is the PESQ, which is defined based on the standard ITU-T P.862 for qualitative evaluations of speech signals in mobile stations [4,42]. In fact, the PESQ criteria is used in the numerical representation of qualitative evaluations for enhanced speech signals. The defined range for this criteria is [−0.5 4.5], where −0.5 and 4.5 show the lowest and highest quality of the enhanced speech, respectively. Additionally, the results were compared with the MOS score criteria. These are qualitative criteria in telecommunication systems that represent the clarity, perception, and intelligibility of the enhanced signal. The MOS criteria are defined based on the standard ITU-T P.800 [5,43] in telecommunication systems. The evaluation results based on the MOS criteria was implemented by the use of some volunteers, by listening to the enhanced signal, where 1 and 5 are the lowest and highest scores in this criteria, respectively. Table 3 shows the defined scores for the MOS criteria in the evaluations.

Table 3. The numerical scores for the mean opinion score (MOS) criteria in the evaluation process.

Rating	Quality (Standard ITU-T P.800)	Impairment
5	Excellent	Imperceptible
4	Good	Perceptible but not annoying
3	Fair	Slightly annoying
2	Poor	Annoying
1	Bad	Very annoying

Finally, the last qualitative criteria for evaluations is the STOI. This criteria predicts the intelligibility of humans based on a series of cases. The speech intelligibility measurement is based on the existence of a series of pre-assumptions, but if the noisy signal is processed based on the time-frequency weighting, the final results are not trustable. The STOI is an objective intelligibility measurement, which represents the highest convolution value by the intelligibility of both noisy and weighted time-frequency noisy signals. In addition, the lowest and highest scores for the STOI criteria are 0 and 1, which represent the best and the worst enhancement performance, respectively.

Firstly, the proposed method was evaluated on the white noise and then, the other colored noise were considered in the experiments. The proposed CNMA-SBAPA was evaluated on real and simulated data in comparison with the LMS, traditional APA, RLS, DB-MWF, and MNMF-MVDR algorithms. Figure 11 shows the time-domain and spectrum for the noisy and enhanced signals in the presence of white noise for SNR = 0 dB. As seen in these figures, the proposed CNMA-SBAPA method decreased more level of the noise with less distortion in comparison with other works. However, the numerical values are necessary for comparison. In the following, the experiments were evaluated with quantitative and qualitative criteria.

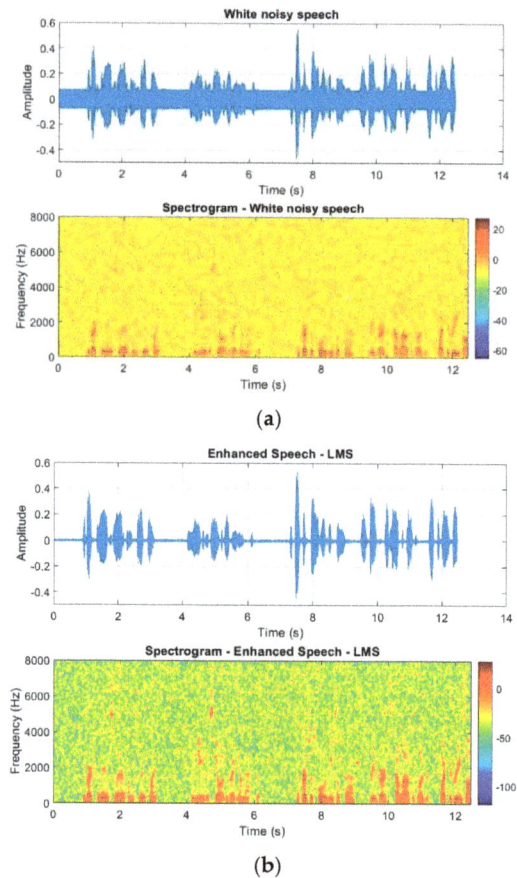

Figure 11. *Cont.*

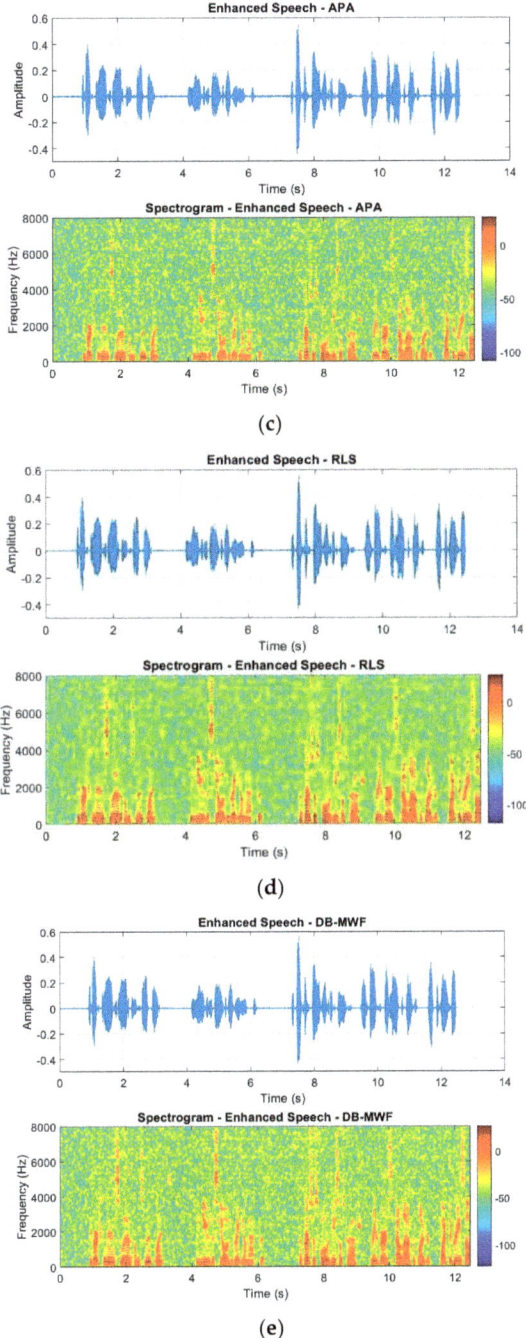

Figure 11. Cont.

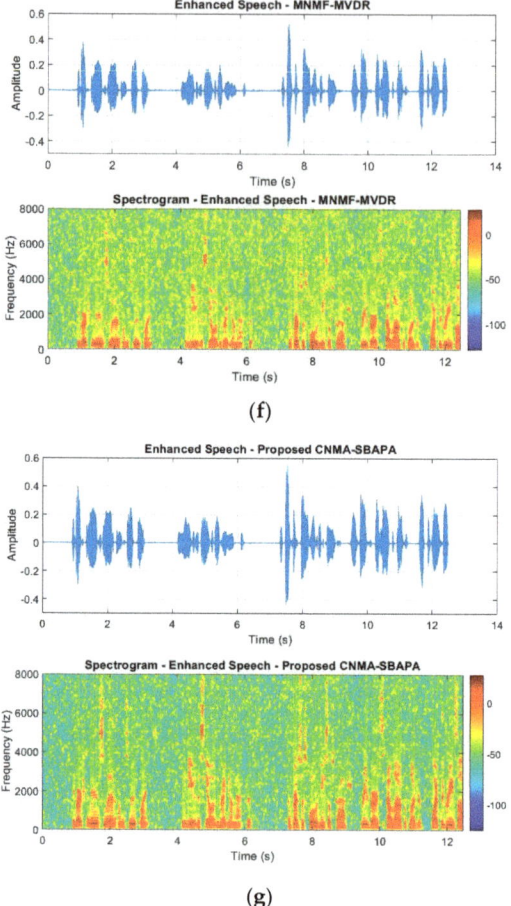

Figure 11. The time-domain and spectrum representation for (**a**) white noisy signal and enhanced signal by the (**b**) least mean square (LMS), (**c**) APA, (**d**) recursive least square (RLS), (**e**)distributed multichannel Wiener filter (DB-MWF), (**f**) multichannel nonnegative matrix factorization-minimum variance distortionless response (MNMF-MVDR), and (**g**) proposed CNMA-SBAPA for SNR=0dB.

In the following, the proposed method was compared by numerical criteria with other previous works. Figure 12 shows the SegSNR results in SNRs [−10, −5, 0, 5, 10, and 15] dB for the proposed CNMA-SBAPA in comparison with the LMS, traditional APA, RLS, DB-MWF, and MNMF-MVDR for real and simulated data in the presence of white noise. As seen, the proposed method had a superior performance in different ranges of SNRs in comparison with the rest of the works, namely a better noise elimination was reached via the proposed algorithm. For example, the proposed method enhanced the noisy speech signal with SNR = −10 dB to SegSNR = 1.35 dB in comparison with SegSNR = −4.58 dB in LMS, SegSNR = −3.21 dB in APA, SegSNR = −1.57 dB in RLS, SegSNR = −1.68 dB in DB-MWF, and SegSNR = −0.94 dB in MNMF-MVDR. Nevertheless, the quantitative criteria are not enough to properly evaluate a method, and both quantitative and qualitative criteria should be considered in the evaluations.

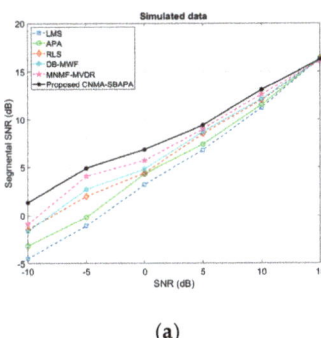 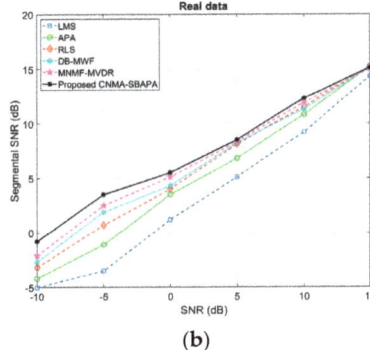

Figure 12. The segmental signal-to-noise ratio (SegSNR) comparison between the proposed CNMA-SBAPA, LMS, traditional APA, RLS, DB-MWF, and MNMF-MVDR methods on (**a**) simulated and (**b**) real data for white noise.

In addition, the proposed method was compared with previous works by qualitative criteria such as the PESQ, MOS, and STOI. We used 20 volunteers, where they listened first to the clean signal by the headset to have an idea of an excellent signal with a rating of 5 in the MOS scale and a noisy signal (before enhancement), which is the worst option in the MOS scale with a rating of 1. Then, the enhanced signal in a different range of SNRs were played for them, and they were asked to select a rate between 1 and 5 based on the Table 3. Figure 13 shows the PESQ, STOI, and averaged MOS criteria for the enhanced signal by the proposed method in comparison with previous works on real and simulated data for different ranges of SNRs in the presence of white noise. As seen, the proposed method had the best performance in comparison with previous works. For example, the PESQ score was 3.41 in the proposed method in comparison to 1.82 in LMS, 2.51 in APA, 2.73 in RLS, 2.93 in DB-MWF, and 3.1 in MNMF-MVDR, for SNR=15dB for simulated data. In addition, the STOI criteria was 0.89 in the proposed method in comparison to 0.73 in LMS, 0.77 in APA, 0.81 in RLS, 0.83 in DB-MWF, and 0.85 in MNMF-MVDR, for SNR=15dB. The other criteria for comparison was the average MOS rate, which was 3.5 in the proposed method in comparison to 2.5 in LMS, 2.7 in APA, 2.9 in RLS, 3.0 in DB-MWF, and 3.0 in MNMF-MVDR, for SNR=15dB. Therefore, the proposed method was superior for enhancing the noisy signals by considering both quantitative (Figure 12) and qualitative (Figure 13) criteria in comparison to previous works in the presence of white noise. In addition, the proposed method was implemented on colored noises to show the reliability of the results. For this purpose, the proposed method was evaluated on babble, train, car, and restaurant noises for the real and simulated data and for SNR ranges [−10,−5,0,5,10, and 15]dB. Tables 4 and 5 show the results on the simulated and real data, respectively. As seen from the numbers in these tables, the proposed method had better results in most cases in comparison with traditional methods, which present the reliability of the proposed method in colored noisy conditions. Some of the methods had slightly better results in specific cases, for example in SNR=15dB, which cannot be generalized to all cases. In addition, the SegSNR values are shown in these tables to present better comparison with qualitative criteria.

Finally, Table 6 presents the speed of convergence for the proposed method in comparison with other previous works for all white and colored noises in seconds (the required time for convergence based on the configuration of the used PC) on the real data. As shown, the proposed method has a higher speed of convergence in comparison with other algorithms. The main reason for this high speed of convergence is the sub-band processing, because this multiresolution processing provides stationary noise in each frequency band, which is an important factor in the speed of convergence. When the noise is closer to stationary conditions, the speed of convergence is increased in adaptive filter-based algorithms. As clearly shown in this table, the speed of convergence in white noisy conditions was higher than the colored noisy scenarios. Therefore, the proposed CNMA-SBAPA method had superiority for the speech enhancement in comparison with LMS, traditional APA, RLS, DB-MWF, and MNMF-MVDR algorithms based on the quantitative SegSNR and qualitative PESQ, MOS, and STOI criteria, as well as the speed of convergence.

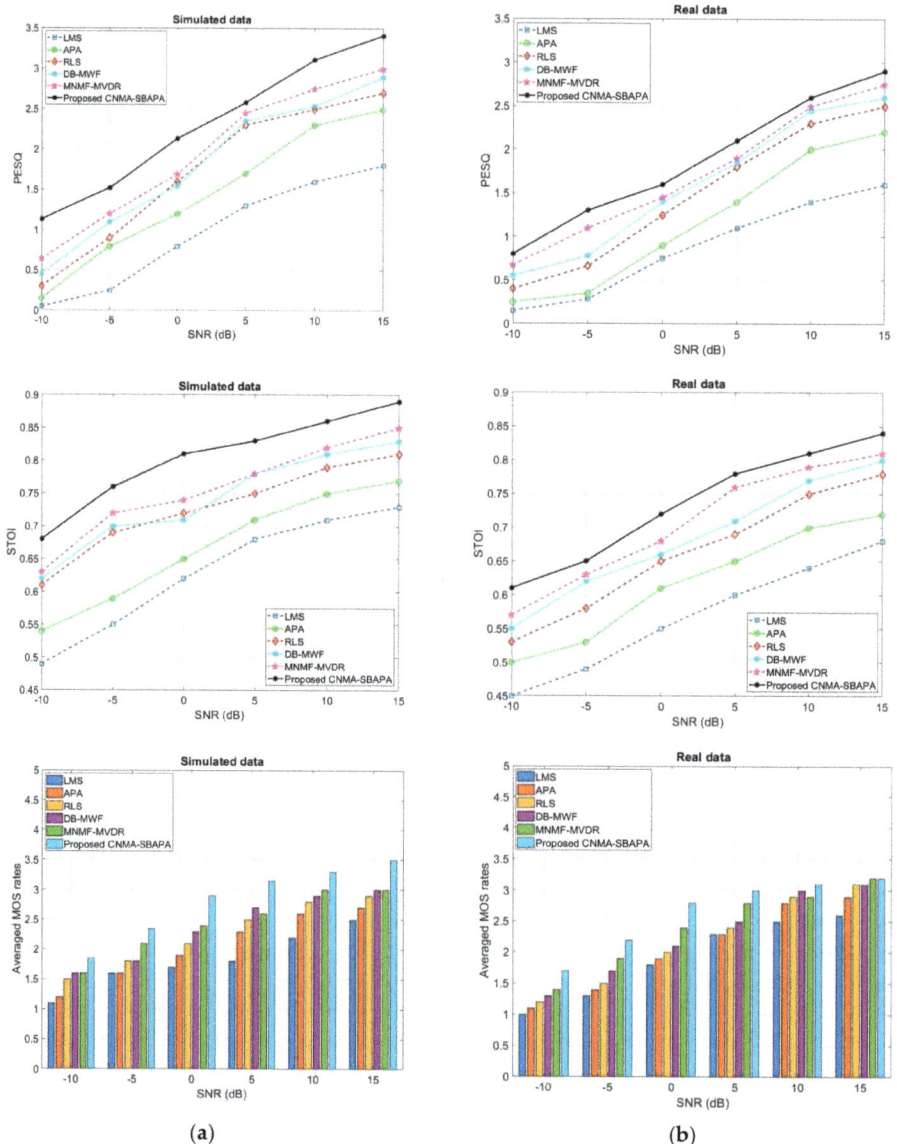

Figure 13. The perceptual evaluation of speech quality (PESQ), short-time objective intelligibility (STOI), and averaged mean opinion score (MOS) comparison between the proposed CNMA-SBAPA and the LMS, traditional APA, RLS, DB-MWF, and MNMF-MVDR methods for (**a**) simulated and (**b**) real data by considering the white noise.

Table 4. The comparison between PESQ, MOS, STOI, and SegSNR for the proposed CNMA-SBAPA in comparison with the LMS, traditional APA, RLS, DB-MWF, and MNMF-MVDR methods on the simulated data for colored noises such as: train, babble, car, and restaurant noises in different range of SNRs (the bold numbers are the best results).

SNR (dB)	Methods	Babble Noise				Train Noise				Car Noise				Restaurant Noise			
		SegSNR	PESQ	STOI	MOS	SegSNR	PESQ	STOI	MOS	SegSNR	PESQ	STOI	MOS	SegSNR	PESQ	STOI	MOS
−10	LMS	−5.23	0.34	0.44	1.10	−6.25	0.29	0.41	1.05	−6.87	0.25	0.36	1.05	−6.87	0.18	0.37	1.00
	APA	−4.63	0.63	0.51	1.15	−5.96	0.56	0.47	1.10	−6.08	0.48	0.43	1.05	−6.08	0.52	0.39	1.10
	RLS	−2.91	0.86	0.58	1.45	−3.98	0.81	0.51	1.35	−4.61	0.74	0.48	1.30	−4.61	0.65	0.44	1.30
	DB-MWF	−2.48	0.92	0.57	1.55	−3.46	0.89	0.53	1.45	−4.77	0.78	0.50	1.45	−4.77	0.79	0.47	1.40
	MNMF-MVDR	−2.23	1.04	0.60	1.60	−3.29	0.96	0.56	1.50	−4.29	0.85	0.53	1.55	−4.29	0.84	0.48	1.45
	CNMA-SBAPA	**−1.71**	**1.19**	**0.65**	**1.85**	**−2.69**	**1.16**	**0.63**	**1.70**	**−3.14**	**1.03**	**0.59**	**1.65**	**−3.14**	**0.99**	**0.56**	**1.60**
−5	LMS	−2.66	0.48	0.53	1.50	−3.59	0.44	0.51	1.40	−3.78	0.39	0.46	1.35	−3.78	0.41	0.47	1.35
	APA	−1.89	0.76	0.59	1.55	−2.41	0.68	0.56	1.45	−3.26	0.61	0.51	1.40	−3.26	0.65	0.54	1.4
	RLS	0.57	0.91	0.65	1.75	−1.14	0.82	0.62	1.70	−1.97	0.79	0.58	1.60	−1.97	0.72	0.60	1.55
	DB-MWF	1.18	0.98	0.67	1.90	−0.35	0.94	0.62	1.80	−0.29	0.87	0.60	1.65	−0.29	0.93	0.59	1.70
	MNMF-MVDR	1.97	1.13	0.68	1.95	0.87	1.06	0.65	1.95	0.96	1.03	0.61	1.75	0.96	0.98	0.61	1.70
	CNMA-SBAPA	**3.63**	**1.43**	**0.73**	**2.30**	**2.73**	**1.36**	**0.70**	**2.20**	**2.26**	**1.32**	**0.68**	**2.15**	**2.26**	**1.20**	**0.64**	**2.10**
0	LMS	3.58	0.75	0.61	1.65	2.69	0.66	0.56	1.55	2.24	0.59	0.52	1.5	2.24	0.54	0.49	1.55
	APA	4.07	1.14	0.63	1.8	3.21	1.05	0.61	1.70	3.56	0.99	0.57	1.65	3.56	0.93	0.55	1.65
	RLS	4.12	1.53	0.71	2.05	3.67	1.47	0.66	1.95	3.92	1.41	0.64	1.9	3.92	1.36	0.62	1.85
	DB-MWF	4.83	1.61	0.72	2.25	3.98	1.59	0.68	2.00	4.11	1.58	0.66	1.95	4.11	1.49	0.65	1.95
	MNMF-MVDR	5.07	1.72	0.75	2.35	4.31	1.67	0.7	2.25	4.39	1.69	0.69	2.15	4.39	1.61	0.65	2.10
	CNMA-SBAPA	**5.40**	**2.04**	**0.78**	**2.85**	**4.95**	**1.98**	**0.75**	**2.70**	**4.78**	**1.91**	**0.73**	**2.65**	**4.78**	**1.84**	**0.75**	**2.55**
5	LMS	8.59	1.22	0.67	1.75	8.25	1.14	0.64	1.70	8.17	1.09	0.60	1.7	8.17	1.01	0.57	1.65
	APA	8.96	1.63	0.69	2.20	8.39	1.57	0.66	2.15	8.22	1.5	0.63	2.05	8.22	1.46	0.64	2.00
	RLS	9.28	2.21	0.76	2.40	8.80	2.12	0.73	2.35	8.64	2.07	0.71	2.35	8.64	1.95	0.67	2.30
	DB-MWF	9.56	2.32	0.75	2.60	9.12	2.08	0.71	2.60	8.82	2.19	0.70	2.40	8.82	2.07	0.66	2.45
	MNMF-MVDR	10.12	2.44	0.78	2.75	9.54	2.34	0.75	2.70	9.25	2.31	0.72	2.60	9.25	2.24	0.68	2.50
	CNMA-SBAPA	**10.63**	**2.59**	**0.82**	**3.25**	**10.21**	**2.53**	**0.80**	**3.10**	**10.03**	**2.48**	**0.78**	**2.95**	**10.03**	**2.49**	**0.73**	**2.90**
10	LMS	12.52	1.53	0.69	2.1	12.11	1.47	0.68	2.05	11.95	1.39	0.65	2.00	11.95	1.41	0.63	1.95
	APA	12.86	2.2	0.73	2.55	12.43	2.13	0.70	2.45	12.27	2.07	0.65	2.40	12.27	2.08	0.66	2.45
	RLS	13.47	2.45	0.78	2.70	12.86	2.39	0.75	2.6	12.54	2.33	0.71	2.55	12.54	2.38	0.70	2.50
	DB-MWF	13.21	2.53	0.78	2.80	12.75	2.48	0.77	2.75	12.56	2.25	0.73	2.65	12.56	2.49	0.72	2.55
	MNMF-MVDR	13.52	2.67	0.79	2.95	12.98	2.61	0.78	2.85	12.71	2.46	0.76	2.70	12.71	2.65	0.75	2.65
	CNMA-SBAPA	**14.02**	**3.08**	**0.83**	**3.20**	**13.28**	**2.96**	**0.82**	**3.10**	**13.09**	**2.87**	**0.79**	**3.05**	**13.09**	**2.92**	**0.79**	**2.95**
15	LMS	15.12	1.76	0.72	2.45	15.13	1.69	0.72	2.35	15.4	1.64	0.69	2.30	15.4	1.68	0.7	2.25
	APA	15.55	2.44	0.76	2.65	15.35	2.37	0.75	2.55	15.32	2.28	0.76	2.50	15.32	2.25	0.74	2.45
	RLS	**15.74**	2.61	0.81	2.85	15.35	2.58	0.79	2.75	15.48	2.51	0.80	2.65	15.48	2.40	0.79	2.65
	DB-MWF	15.52	2.99	0.83	3.00	15.24	2.83	0.76	2.90	**15.52**	2.72	0.79	2.75	**15.52**	2.71	0.80	2.80
	MNMF-MVDR	15.63	3.03	0.84	3.10	15.42	2.96	0.78	3.05	15.45	2.93	**0.81**	2.90	15.45	2.80	**0.83**	2.85
	CNMA-SBAPA	15.71	**3.31**	**0.89**	**3.35**	**15.69**	**3.22**	**0.84**	**3.30**	15.44	**3.24**	0.80	**3.25**	15.44	**3.19**	0.82	**3.20**

Table 5. The comparison between PESQ, MOS, STOI, and SegSNR for the proposed CNMA-SBAPA in comparison with the LMS, traditional APA, RLS, DB-MWF, and MNMF-MVDR methods on the real data for colored noises such as: train, babble, car, and restaurant noises in different range of SNRs (the bold numbers are the best results).

SNR (dB)	Methods	Babble Noise				Train Noise				Car Noise				Restaurant Noise			
		SegSNR	PESQ	STOI	MOS	SegSNR	PESQ	STOI	MOS	SegSNR	PESQ	STOI	MOS	SegSNR	PESQ	STOI	MOS
−10	LMS	−5.56	0.32	0.41	1.00	−5.82	0.25	0.39	1.00	−6.46	0.18	0.36	1.00	−6.92	0.14	0.35	1.00
	APA	−4.82	0.61	0.47	1.10	−5.42	0.55	0.46	1.15	−5.83	0.51	0.42	1.10	−6.21	0.48	0.39	1.05
	RLS	−3.04	0.82	0.52	1.35	−3.87	0.74	0.51	1.30	−4.52	0.71	0.49	1.20	−4.86	0.64	0.45	1.15
	DB-MWF	−2.98	0.86	0.55	1.40	−3.43	0.76	0.53	1.35	−4.13	0.75	0.50	1.30	−4.51	0.67	0.45	1.25
	MNMF-MVDR	−2.54	0.93	0.57	1.50	−3.01	0.83	0.52	1.45	−3.81	0.80	0.51	1.35	−4.09	0.75	0.47	1.30
	CNMA-SBAPA	−2.21	1.08	0.58	1.65	−2.67	1.01	0.56	1.50	−3.14	0.96	0.53	1.45	−3.46	0.92	0.53	1.40
−5	LMS	−2.83	0.46	0.50	1.45	−3.2	0.44	0.48	1.40	−3.54	0.41	0.47	1.30	−3.92	0.38	0.43	1.25
	APA	−2.04	0.72	0.55	1.50	−2.57	0.68	0.53	1.45	−3.09	0.64	0.52	1.45	−3.47	0.61	0.49	1.35
	RLS	0.14	0.88	0.62	1.60	−0.52	0.82	0.61	1.55	−1.45	0.74	0.58	1.50	−1.88	0.69	0.54	1.50
	DB-MWF	0.93	0.91	0.64	1.75	0.12	0.85	0.64	1.65	−0.78	0.79	0.60	1.60	−1.57	0.73	0.56	1.65
	MNMF-MVDR	1.53	1.02	0.67	1.90	0.84	0.97	0.65	1.70	0.21	0.93	0.59	1.75	−0.84	0.88	0.59	1.75
	CNMA-SBAPA	3.29	1.29	0.70	2.10	2.76	1.21	0.67	2.05	2.25	1.16	0.64	1.95	1.76	1.11	0.61	1.90
0	LMS	3.25	0.72	0.57	1.5	2.84	0.65	0.52	1.45	2.43	0.59	0.51	1.35	2.07	0.53	0.46	1.30
	APA	3.99	1.05	0.62	1.75	3.71	1.01	0.57	1.65	3.67	0.94	0.52	1.6	3.44	0.89	0.51	1.50
	RLS	4.15	1.49	0.69	1.95	4.03	1.44	0.65	1.85	3.95	1.38	0.62	1.8	3.79	1.33	0.59	1.70
	DB-MWF	4.32	1.54	0.7	2.05	4.26	1.51	0.67	1.95	4.04	1.44	0.64	1.9	3.93	1.40	0.60	1.75
	MNMF-MVDR	4.51	1.69	0.72	2.20	4.49	1.58	0.67	2.05	4.29	1.56	0.65	2.00	4.17	1.58	0.63	1.95
	CNMA-SBAPA	5.03	1.95	0.74	2.55	4.86	1.87	0.71	2.50	4.71	1.76	0.70	2.45	4.58	1.72	0.67	2.45
5	LMS	8.41	1.18	0.66	1.6	8.22	1.09	0.62	1.55	8.14	1.02	0.59	1.50	8.01	0.96	0.56	1.50
	APA	8.73	1.58	0.65	2.05	8.52	1.52	0.63	2.00	8.39	1.48	0.60	1.95	8.25	1.41	0.61	1.90
	RLS	9.04	2.15	0.74	2.25	8.96	2.04	0.71	2.2	8.71	1.96	0.70	2.15	8.52	1.89	0.66	2.05
	DB-MWF	9.32	2.23	0.73	2.5	9.38	2.05	0.7	2.4	8.83	2.06	0.70	2.25	8.86	1.92	0.65	2.10
	MNMF-MVDR	9.58	2.34	0.76	2.65	9.62	2.26	0.72	2.55	9.22	2.19	0.71	2.40	9.32	2.08	0.66	2.35
	CNMA-SBAPA	10.27	2.48	0.78	3.00	10.08	2.41	0.75	2.90	9.95	2.36	0.72	2.85	9.76	2.32	0.69	2.75
10	LMS	12.43	1.48	0.67	2.05	12.26	1.45	0.66	1.95	12.09	1.40	0.64	1.90	11.87	1.37	0.65	1.95
	APA	12.91	2.15	0.70	2.40	12.62	2.11	0.69	2.35	12.47	2.07	0.67	2.35	12.31	2.01	0.66	2.25
	RLS	13.28	2.41	0.75	2.55	13.01	2.39	0.74	2.50	12.76	2.36	0.72	2.40	12.62	2.31	0.73	2.35
	DB-MWF	13.36	2.48	0.76	2.75	13.22	2.49	0.73	2.60	12.89	2.47	0.71	2.50	12.75	2.38	0.74	2.45
	MNMF-MVDR	13.51	2.56	0.76	2.90	13.34	2.55	0.75	2.75	12.96	2.51	0.74	2.65	12.83	2.48	0.72	2.70
	CNMA-SBAPA	13.88	2.92	0.81	3.05	13.51	2.88	0.79	3.00	13.19	2.82	0.77	2.95	12.99	2.79	0.75	3.00
15	LMS	15.08	1.74	0.70	2.40	15.1	1.69	0.67	2.30	15.04	1.65	0.67	2.15	15.25	1.60	0.66	2.10
	APA	15.41	2.38	0.73	2.55	15.38	2.35	0.72	2.50	15.86	2.28	0.70	2.40	15.45	2.21	0.7	2.45
	RLS	15.24	2.56	0.84	2.70	15.29	2.51	0.77	2.70	15.92	2.44	0.75	2.60	15.76	2.37	0.74	2.55
	DB-MWF	15.39	2.81	0.85	2.85	15.36	2.62	0.77	2.80	15.81	2.50	0.78	2.75	15.78	2.46	0.77	2.65
	MNMF-MVDR	15.48	2.94	0.82	3.00	15.42	2.74	0.79	2.90	15.80	2.68	0.80	2.85	15.73	2.59	0.74	2.70
	CNMA-SBAPA	15.35	3.22	0.83	3.20	15.49	3.15	0.82	3.25	15.83	3.13	0.77	3.15	15.61	3.09	0.73	3.10

Table 6. The speed of convergence, in seconds, for the proposed CNMA-SBAPA in comparison with the LMS, traditional APA, RLS, DB-MWF, and MNMF-MVDR methods for white and colored noises on real data in different range of SNRs (the bold numbers are the best results).

SNR(dB)	Methods	Speed of Convergence (Seconds)				
		White Noise	Babble Noise	Train Noise	Car Noise	Restaurant Noise
−10	LMS	0.541	0.582	0.61	0.654	0.668
	APA	0.516	0.551	0.592	0.611	0.627
	RLS	0.422	0.468	0.496	0.539	0.546
	DB-MWF	0.586	0.612	0.652	0.673	0.695
	MNMF-MVDR	0.51	0.539	0.57	0.592	0.637
	CNMA-SBAPA	**0.356**	**0.367**	**0.393**	**0.419**	**0.427**
−5	LMS	0.537	0.556	0.579	0.601	0.634
	APA	0.497	0.527	0.541	0.56	0.572
	RLS	0.403	0.429	0.447	0.482	0.506
	DB-MWF	0.545	0.593	0.615	0.636	0.658
	MNMF-MVDR	0.494	0.509	0.527	0.553	0.587
	CNMA-SBAPA	**0.337**	**0.356**	**0.379**	**0.391**	**0.411**
0	LMS	0.516	0.538	0.562	0.568	0.595
	APA	0.473	0.482	0.502	0.536	0.539
	RLS	0.396	0.409	0.427	0.435	0.452
	DB-MWF	0.531	0.563	0.579	0.602	0.621
	MNMF-MVDR	0.485	0.498	0.516	0.531	0.546
	CNMA-SBAPA	**0.318**	**0.329**	**0.35**	**0.358**	**0.362**
5	LMS	0.492	0.505	0.517	0.525	0.529
	APA	0.464	0.479	0.492	0.467	0.503
	RLS	0.388	0.395	0.401	0.412	0.418
	DB-MWF	0.507	0.512	0.544	0.565	0.586
	MNMF-MVDR	0.459	0.466	0.487	0.503	0.525
	CNMA-SBAPA	**0.327**	**0.336**	**0.347**	**0.359**	**0.365**
10	LMS	0.488	0.498	0.509	0.521	0.538
	APA	0.451	0.467	0.483	0.499	0.507
	RLS	0.369	0.381	0.389	0.395	0.411
	DB-MWF	0.478	0.496	0.513	0.543	0.559
	MNMF-MVDR	0.437	0.458	0.479	0.491	0.516
	CNMA-SBAPA	**0.305**	**0.328**	**0.339**	**0.352**	**0.368**
15	LMS	0.472	0.485	0.493	0.498	0.506
	APA	0.463	0.474	0.478	0.485	0.49
	RLS	0.372	0.376	0.385	0.396	0.408
	DB-MWF	0.443	0.455	0.462	0.481	0.494
	MNMF-MVDR	0.41	0.427	0.448	0.463	0.48
	CNMA-SBAPA	**0.299**	**0.319**	**0.325**	**0.349**	**0.374**

5. Conclusions

Speech enhancement is an important application in the signal processing for smart meeting rooms. The aim of speech enhancement is denoising, dereverberation, or denoising–dereverberation at the same time. The speech enhancement is implemented as a pre-processing step to produce the proper signal in such an application as speaker localization, tracking, speech recognition, text-to-speech, estimation the number of speakers, etc. The speech enhancement algorithms are divided into the single and multi-channels methods. The single-channel algorithms are challenging in the speech enhancement processes because of the lack of suitable information in the denoising procedure. In contrast, the multi-channel algorithms increase the enhancement accuracy due to having more information but the computational complexity is increased. In this article, a multi-channel speech enhancement method was proposed based on the microphone array. The microphone array increased the accuracy in the enhanced algorithms based on the increasing of information, but the spatial aliasing decreased the efficiency because of inter-microphone distances. In this article, a uniform circular nested microphone array was proposed for the speech enhancement algorithms. This nested array was designed in a way that the microphones were located at specific distances to eliminate the spatial aliasing, in combination with analysis filters to provide the proper information for the speech enhancement algorithms. In addition, the speech information is different in various frequency bands. Therefore, the specific sub-band processing was proposed to have especial attention to the speech spectrum components. The frequency bands were designed to have the maximum resolution in low frequency components. In the following, the APA was implemented on all frequency bands, which was obtained by the sub-band processing and circular nested microphone array. The projection factor ($N=4$) was considered for the CNMA-SBAPA in order to keep the computational complexity in an acceptable range along with the superior accuracy. Finally, the synthesis filter bank was implemented on the sub-band signals and the enhanced signal was generated by the summation through all sub-bands. The proposed algorithm was compared with the LMS, traditional APA, RLS, DB-MWF, and MNMF-MVDR methods on the real and simulated data for white and colored noises under the SNRs range [−10,−5,0,5,10, and 15]dB. In all conditions the proposed method had a superior accuracy in comparison with previous works. In addition, the proposed method was compared based on the speed of convergence with previous works, which it was much faster among all the other algorithms. Since the proposed enhancement algorithm was implemented on stationary signals, where its benefit was increasing the speed of convergence in adaptive filters.

One of the future works is reducing the size of the array and decreasing the number of microphones (without having a high effect on the quality) to be applicable for smartphone applications. Even the type of the microphones is important. In this article, we used a high quality microphone, which provides the signals with proper amplitude from the environment. The use of normal microphones in smartphones is another challenge, which could be an area for future work. Another area for future work is to find the best numbers of sub-bands to provide the maximum performance and lowest computational complexity, where the numbers of sub-bands will not be fixed and it should be adaptive based on the speech components.

Author Contributions: Conceptualization, A.D.F. and P.A. and D.Z.-B.; methodology, A.D.F. and P.A.; software, A.D.F., P.I., P.A. and H.D.; validation, M.S., P.P., D.Z.-B. and C.A.-M.; formal analysis, A.D.F. and P.A.; investigation, A.D.F. and P.A.; resources, A.D.F., P.A., D.Z.-B. and P.I.; data curation, A.D.F.; writing—original draft preparation, A.D.F., P.A. and D.Z.-B.; writing—review and editing, P.P.-J., C.A. and D.Z.-B.; supervision, P.I.; project administration, P.A., H.D., M.S. and D.Z.-B.; funding acquisition, P.A. and A.D.F. All authors have read and agreed to the published version of the manuscript.

Funding: This research was funded by FONDECYT Postdoctorado No. 3190147, FONDECYT No. 11180107 and ANID PFCHA/Beca de Doctorado Nacional/2019 21190489.

Acknowledgments: This work was supported by the Vicerrectoría de Investigación y Postgrado of the Universidad Tecnológica Metropolitana, the Vicerrectoría de Investigación y Postgrado, and Faculty of Engineering Science of the Universidad Católica del Maule.

Conflicts of Interest: The authors declare no conflict of interest.

Abbreviations

The following abbreviations are used in this manuscript:

AP	Affine projection
APA	Affine projection algorithm
AR	Auto-regressive
ARHMM	Auto-regressive hidden Markov model
CNMA	Circular nested microphone array
CNMA-SBAPA	Circular nested microphone array in combination with sub-band affine projection algorithm

DB-MWF	Distributed multichannel Wiener filter
DNN	Deep neural network
FBK	Fondazione Bruno Kessler
FIR	Finite impulse response
HMM	Hidden Markov model
HPF	High-pass filter
IFD	Instantaneous frequency deviation
LMR-APA	Levenberg Marquardt regularized-Affine projection algorithm
LMS	Least mean square
LPF	Low-pass filter
ML	Maximum likelihood
MSE	Mean square error
MMSE	Minimum mean square error
MNMF	Multi-channel non-negative matrix factorization
MNMF-MVDR	Multichannel nonnegative matrix factorization-minimum variance distortionless response
MOS	Mean opinion score
MVDR	Minimum variance distortionless response
NLMS	Normalized least mean square
NMA	Nested microphone array
OCF-NLMS	Orthogonal correction factor-Normalized least mean square
PESQ	Perceptual evaluation of speech quality
PRAPA	Partial rank affine projection algorithm
RLS	Recursive least square
SBAPA	Sub-band affine projection algorithm
SCM	Spatial covariance matrix
SegSNR	Segmental signal-to-noise ratio
SNR	Signal-to-noise ratio
STOI	Short-time objective intelligibility
STP	Short time predictor
TTS	Text-to-speech
VAD	Voice activity detector
WF	Wiener filter

References

1. Prasad, P.B.M.; Ganesh, M.S.; Gangashetty, S.V. Two microphone technique to improve the speech intelligibility under noisy environment. In Proceedings of the IEEE 14th International Colloquium on Signal Processing & Its Applications (CSPA), Penang, Malaysia, 9–10 March 2018; pp. 13–18.
2. Fukui, M.; Shimauchi, S.; Hioka, Y.; Nakagawa, A.; Haneda, Y. Acoustic echo and noise canceller for personal hands-free video IP phone. *IEEE Trans. Consum. Electron.* **2016**, *62*, 454–462. [CrossRef]
3. Ephraim, Y. Statistical-Model-Based Speech Enhancement Systems. *Proc. IEEE.* **1992**, *80*, 1526–1555. [CrossRef]
4. Rix, A.W.; Beerends, J.G.; Hollier, M.P.; Hekstra, A.P. Perceptual evaluation of speech quality (PESQ)—A new method for speech quality assessment of telephone net works and codecs. In Proceedings of the IEEE International Conference on Acoustics, Speech, and Signal Processing (Cat. No. 01CH37221), Salt Lake City, UT, USA, 7–11 May 2001; pp. 749–752.
5. Streijl, R.C.; Winkler, S.; Hands, D.S. Mean opinion score (MOS) revisited: Methods and applications, limitations and alternatives. *Multimed. Syst.* **2016**, *22*, 213–227. [CrossRef]
6. Taal, C.H.; Hendriks, R.C.; Heusdens, R.; Jensen, J. A short-time objective intelligibility measure for time-frequency weighted noisy speech. In Proceedings of the IEEE International Conference on Acoustics, Speech and Signal Processing, Dallas, TX, USA, 14–19 March 2010; pp. 4214–4217.
7. Pollak, P.; Vondrasek, M. Methods for Speech SNR Estimation: Evaluation Tool and Analysis of VAD Dependency. *Radio Eng.* **2005**, *14*, 6–11.

8. Loizou, P.C. *Speech Enhancement: Theory and Practice*, 2nd ed.; CRC Press: Boca Raton, FL, USA, 2007.
9. Doclo, S.; Moonen, M.; Bogaert, T.V.; Wouters, J. Reduced-band width and distributed MWF-based noise reduction algorithms for binaural hearing aids. *IEEE Trans. Audio Speech Lang. Process.* **2009**, *17*, 38–51. [CrossRef]
10. Boll, S.F. Suppression of Acoustic Noise in Speech Using Spectral Subtraction. *IEEE Trans. Acoust. Speech Signal Process.* **1979**, *27*, 113–120. [CrossRef]
11. Martin, R. Spectral subtraction based on minimum statistics. In Proceedings of the European Signal Processing Conference, Scotland, UK, 13–16 September 1994; pp. 1182–1185.
12. Ephraim, Y.; Malah, D. Speech enhancement using a minimum mean square error short-time spectral amplitude estimator. *IEEE Trans. Acoust. Speech Signal Process.* **1984**, *32*, 1109–1121. [CrossRef]
13. Ephraim, Y.; Malah, D. Speech enhancement using a minimum mean-square error log-spectral amplitude estimator. *IEEE Trans. Acoust. Speech Signal Process.* **1985**, *33*, 443–445. [CrossRef]
14. Cohen, I.; Berdugo, B. Noise estimation by minima controlled recursive averaging for robust speech enhancement. *IEEE Signal Process. Lett.* **2002**, *9*, 12–15. [CrossRef]
15. Cohen, I. Noise spectrum estimation in adverse environments: Improved minima controlled recursive averaging. *IEEE Trans. Speech Audio Process.* **2003**, *11*, 466–475. [CrossRef]
16. Martin, R. Noise power spectral density estimation based on optimal smoothing and minimum statistics. *IEEE Trans. Speech Audio Process.* **2001**, *9*, 504–512. [CrossRef]
17. Sameti, H.; Sheikhzadeh, H.; Deng, L.; Brennan, R.L. HMM-based strategies for enhancement of speech signals embedded in non stationary noise. *IEEE Trans. Speech Audio Process.* **1998**, *6*, 445–455. [CrossRef]
18. Zhao, D.Y.; Kleijn, W.B. HMM-based gain modeling for enhancement of speech in noise. *IEEE Trans. Audio Speech Lang. Process.* **2007**, *15*, 882–892. [CrossRef]
19. Deng, F.; Bao, C.C.; Kleijin, W.B. Sparse Hidden Markov Models for Speech Enhancement in Non-Stationary Noise Environments. *IEEE Trans. Audio Speech Lang. Process.* **2015**, *23*, 1973–1987. [CrossRef]
20. Geravanchizadeh, M.; Osgouei, S.G. Dual-channel speech enhancement using normalized fractional least-mean-squares algorithm. In Proceedings of the 19th Iranian Conference on Electrical Engineering, Tehran, Iran, 17–19 May 2011; pp. 1–5.
21. Rakesh, P.; Kumar, T.K. A novel RLS based adaptive filtering method for speech enhancement. *Int. J. Electr. Comput. Electron. Commun. Eng.* **2015**, *9*, 153–158.
22. He, Q.; Bao, F.; Bao, C. Multiplicative Update of Auto-Regressive Gains for Codebook-Based Speech Enhancement. *IEEE/ACM Trans. Audio Speech Lang. Process.* **2017**, *25*, 457–468. [CrossRef]
23. Tavakoli, V.M.; Jensen, J.R.; Christensen, M.G.; Benesty, J. A Framework for Speech Enhancement with Ad Hoc Microphone Arrays. *IEEE/ACM Trans. Audio Speech Lang. Process.* **2016**, *24*, 1038–1051. [CrossRef]
24. Shimada, K.; Bando, Y.; Mimura, M.; Itoyama, K.; Yoshii, K.; Kawahara, T. Unsupervised Speech Enhancement Based on Multi channel NMF-Informed Beamforming for Noise-Robust Automatic Speech Recognition. *IEEE/ACM Trans. Audio Speech Lang. Process.* **2019**, *27*, 960–971. [CrossRef]
25. Kavalekalam, M.S.; Nielsen, J.K.; Boldt, J.B.; Christensen, M.G. Model-Based Speech Enhancement for Intelligibility Improvement in Binaural Hearing Aids. *IEEE/ACM Trans. Audio Speech Lang. Process.* **2019**, *27*, 99–113. [CrossRef]
26. Valentini-Botinhao, C.; Yamagishi, J. Speech Enhancement of Noisy and Reverberant Speech for Text-to-Speech. *IEEE/ACM Trans. Audio Speech Lang. Process.* **2018**, *26*, 1420–1433. [CrossRef]
27. Wang, Y.; Brookes, M. Model-Based Speech Enhancement in the Modulation Domain. *IEEE/ACM Trans. Audio Speech Lang. Process.* **2018**, *26*, 580–594. [CrossRef]
28. Koutrouvelis, A.I.; Hendriks, R.C.; Heusdens, R.; Jensen, J. Robust Joint Estimation of Multi microphone Signal Model Parameters. *IEEE/ACM Trans. Audio Speech Lang. Process.* **2019**, *27*, 1136–1150. [CrossRef]
29. Zheng, Y.R.; Goubran, R.A.; El-Tanany, M. Experimental evaluation of a nested microphone array with adaptive noise cancellers. *IEEE Trans. Instrum. Meas.* **2004**, *53*, 777–786. [CrossRef]
30. Haykin, S. *Adaptive Filter Theory*, 4th ed.; Prentice-Hall: Upper Saddle River, NJ, USA, 2002.
31. Gonzalez, A.; Ferrer, M.; Albu, F.; Diego, M. Affine projection algorithms: Evolution to smart and fast algorithms and applications. In Proceedings of the 20th European Signal Processing Conference (EUSIPCO), Bucharest, Romania, 27–31 August 2012; pp. 1965–1969.

32. Sankaran, S.G.; Beex, A.A.L. Normalized LMS algorithm with orthogonal correction factors. In Proceedings of the Conference Record of the Thirty-First Asilomar Conference on Signals, Systems and Computers, Pacific Grove, CA, USA, 2–5 November 1997; pp. 1670–1673.
33. Kratzer, S.G.; Morgan, D.R. The partial Rank Algorithm for adaptive beamforming. In Proceedings of the SPIE0564, Real-Time Signal Processing VIII, San Diego, CA, USA, 22–23 August 1985; pp. 9–14.
34. Ozeki, K.; Umeda, T. An adaptive filtering algorithm using an orthogonal projection to an affine subspace and its properties. *Electron. Commun. Jpn.* **1984**, *67-A*, 19–27. [CrossRef]
35. Gay, S.L.; Benesty, J. *Acoustic Signal Processing for Telecommunication*, 2nd ed.; Springer: Boston, MA, USA, 2000.
36. Waterschoot, T.V.; Rombouts, G.; Moonen, M. Optimally regularized adaptive filtering algorithms for room acoustic signal enhancement. *Signal Process.* **2008**, *88*, 594–611. [CrossRef]
37. Gabrea, M. Double affine projection algorithm-based speech enhancement algorithm. In Proceedings of the IEEE International Conference on Acoustics, Speech, and Signal Processing (ICASSP), Hong Kong, China, 6–10 April 2003; pp. I904–I907.
38. Shin, H.C.; Sayed, A.H.; Song, W.J. Variable step-size NLMS and affine projection algorithms. *IEEE Signal Process Lett.* **2004**, *11*, 132–135. [CrossRef]
39. Garofolo, J.S.; Lamel, L.F.; Fisher, W.M.; Fiscus, J.G.; Pallett, D.S.; Dahlgren, N.L.; Zue, V. TIMIT Acoustic-Phonetic Continuous Speech Corpus LDC93S1. Web Download. Philadelphia: Linguistic Data Consortium, 1993. Available online: https://catalog.ldc.upenn.edu/LDC93S1 (accessed on March 2019).
40. Schwartz, O.; David, A.; Shahen-Tov, O.; Gannot, S. Multi-microphone voice activity and single-talk detectors based on steered-response power output entropy. In Proceedings of the IEEE International Conference on the Science of Electrical Engineering in Israel (ICSEE), Eilat, Israel, 12–14 December 2018; pp. 1–4.
41. Allen, J.; Berkley, D. Image method for efficiently simulating small-room acoustics. *J. Acoust. Soc. Am.* **1979**, *65*, 943–950. [CrossRef]
42. *ITU-T: Methods for Subjective Determination of Transmission Quality*; Recommendation P.862; International Telecommunications Union (ITU-T): Place des Nations, Geneva, Switzerland, 1996.
43. *ITU-T: Methods for Subjective Determination of Transmission Quality*; Recommendation P.800; International Telecommunications Union (ITU-T): Place des Nations, Geneva, Switzerland, 1996.

© 2020 by the authors. Licensee MDPI, Basel, Switzerland. This article is an open access article distributed under the terms and conditions of the Creative Commons Attribution (CC BY) license (http://creativecommons.org/licenses/by/4.0/).

Article

Classification of Hydroacoustic Signals Based on Harmonic Wavelets and a Deep Learning Artificial Intelligence System

Dmitry Kaplun [1], Alexander Voznesensky [1], Sergei Romanov [1], Valery Andreev [2] and Denis Butusov [3],*

1. Department of Automation and Control Processes, Saint Petersburg Electrotechnical University "LETI", Saint Petersburg 197376, Russia; dikaplun@etu.ru (D.K.); a-voznesensky@yandex.ru (A.V.); saromanov@etu.ru (S.R.)
2. Department of Computer-Aided Design, Saint Petersburg Electrotechnical University "LETI", Saint Petersburg 197376, Russia; vsandreev@etu.ru
3. Youth Research Institute, Saint Petersburg Electrotechnical University "LETI", Saint Petersburg 197376, Russia
* Correspondence: dnbutusov@etu.ru; Tel.: +7-950-008-7190

Received: 10 April 2020; Accepted: 26 April 2020; Published: 29 April 2020

Abstract: This paper considers two approaches to hydroacoustic signal classification, taking the sounds made by whales as an example: a method based on harmonic wavelets and a technique involving deep learning neural networks. The study deals with the classification of hydroacoustic signals using coefficients of the harmonic wavelet transform (fast computation), short-time Fourier transform (spectrogram) and Fourier transform using a kNN-algorithm. Classification quality metrics (precision, recall and accuracy) are given for different signal-to-noise ratios. ROC curves were also obtained. The use of the deep neural network for classification of whales' sounds is considered. The effectiveness of using harmonic wavelets for the classification of complex non-stationary signals is proved. A technique to reduce the feature space dimension using a 'modulo N reduction' method is proposed. A classification of 26 individual whales from the Whale FM Project dataset is presented. It is shown that the deep-learning-based approach provides the best result for the Whale FM Project dataset both for whale types and individuals.

Keywords: harmonic wavelets; classification; kNN-algorithm; deep neural networks; machine learning; Fourier transform; short-time Fourier transform; wavelet transform; spectrogram; confusion matrix; ROC curve

---

1. Introduction

The whale was one of the main commercial animals in the past. Whalers were attracted by the huge carcass of this animal—from one whale they could get much more fat and meat than from any other marine animal. Today, many of its species have almost been driven to extinction. For this reason, they are listed in the IUCN Red List of Threatened Species [1]. Currently, the main threat to whales is an anthropogenic factor, expressed in violation of their usual way of life and pollution of the seas. To ensure the safety of rare animals, the number of individuals must be monitored. Within the framework of environmental monitoring programs approved by governments and public organizations of different countries, cetacean monitoring activities are carried out year-round using all of the modern achievements in data processing [2]. Monitoring includes work at sea and post-processing of the collected data: determining the coordinates of whale encounters, establishing the composition of the group, and photographing the animals for subsequent observation of individually recognizable individuals.

Systematic observation of animals presents scientists with the opportunity to learn about how mammals share the water area among themselves, to collect data on age and gender composition [3]. An important task is to find out where the whales come from and where they then go to in the winter, to track their routes of movements. You must also be able to determine which population the whales belong to.

Sounds made by cetaceans for communication are called "whale songs". The word "songs" is used to emphasize the repeating and melodic nature of these sounds, reminiscent of human singing. The use of sounds as the main communication channel is due to the fact that, in an aquatic environment, visibility can be limited, and smells spread much slower than in air [4]. It is believed that the most complex songs of humpback whales and some toothless whales are used in mating games. Simpler signals are used all year round and perhaps serve for day-to-day communication and navigation. Toothed whales (including killer whales) use emitted sounds for echolocation. In addition, it was found that whales that have lived in captivity for long can mimic human speech. All these signals are transmitted to different distances, under different water conditions and in the presence of a variety of noises. Additionally, stable flocks have their own dialects, i.e., there is wide variability in the sounds made by whales, both within the population and between populations. Thus, sounds can be used to classify both whale species and individuals. The task of classifying whales by sound has been solved by many researchers for different types of whales in different parts of the world, using various methods and approaches, the most popular being signal processing algorithms [5,6] and algorithms based on neural networks [2,7–10]. Neural-network-based approaches present different architectures, models and learning methods. In [2], the authors developed and empirically studied a variety of deep neural networks to detect the vocalizations of endangered North Atlantic right whales. In [7], an effective data-driven approach based on pre-trained convolutional neural networks (CNN) using multi-scale waveforms and time-frequency feature representations was developed in order to perform classification of whale calls from a large open-source dataset recorded by sensors carried by whales. The authors of [8] constructed an ensembled deep learning CNN model to classify beluga detections. The applicability of basic CNN models is also being explored for the bio-acoustic task of whale call detection, such as with respect to North Atlantic right whale calls [9] and humpback whale calls [10].

This paper considers two approaches to hydroacoustic classification, taking the sounds made by whales as examples: on the basis of harmonic wavelets and deep learning neural networks. The main contributions of our work can be summarized as follows. The effectiveness of using harmonic wavelets for the classification of hydroacoustic signals was proved. A technique to reduce the feature space dimension using a 'modulo N reduction' method was developed. A classification of 26 individual whales is presented for the dataset. It was shown that the deep-learning-based approach provides the best result for the dataset both for whale types and individuals.

The remainder of this paper is organized as follows. In Section 2, we briefly describe hydroacoustic signal processing and review related works on it. In Section 3, we introduce details of the harmonic wavelets and their application to the processing of hydroacoustic signals. In Section 4, we review the kNN algorithm for classification based on harmonic wavelets and present experimental results to verify the proposed approach. In Section 5, experimental results are presented to verify the approach for classification based on neural networks and machine learning. In Section 6, we discuss the results and how they can be interpreted from the perspective of previous studies and of the working hypotheses. Future research directions also are highlighted. Finally, we present the conclusions in Section 7.

2. Hydroacoustic Signal Processing

Before classifying hydroacoustic signals, which are sounds made by whales in an aquatic environment, they must be pre-processed, as the quality of the classification will depend on their quality. Hydroacoustic signal processing includes data preparation, as well as the use of further algorithms allowing the extraction of useful signals from certain directions. Preliminary processing includes de-noising, estimation of the degree of randomness, extraction of short-term local features, pre-filtering, etc. Preprocessing affects the process of further analysis within a hydroacoustic monitoring system [11–13]. Even though the preprocessing of hydroacoustic signals has been studied for a long time, there are several unresolved problems, namely: working in conditions of a priori uncertainty of signal parameters; processing complex non-stationary hydroacoustic signals with multiple local features; and analysis of multicomponent signals. Another set of problems is represented by effective preliminary visual processing of hydroacoustic signals and the need for a mathematical apparatus for signal preprocessing tasks.

Current advances in applied mathematics and digital signal processing along with the development of high-performance hardware allow the effective application of numerous mathematical techniques, including continuous and discrete wavelet transforms. Wavelets are an effective tool for signal preprocessing, due to their adaptability, the availability of fast computational algorithms and the diversity of wavelet bases.

Using wavelets for hydroacoustic signal analysis provides the following possibilities [14,15]:

1. Detection of foreign objects in marine and river areas, including icebergs and other ice formations, the size estimation of these objects, hazard assessment based on analyzing local signal features;
2. Detection and classification of marine targets based on the analysis of local signal features;
3. Detection of hydroacoustic signals in the presence of background noise;
4. Efficient visualization and processing of hydroacoustic signals based on multiscale wavelet spectrograms.

Classification is an important task of modern signal processing. The quality of the classification depends on the noise level, training size and testing datasets, and the algorithm. It is also important to choose classification features and determine the size of the feature space. The classification feature is the feature or characteristic of the object used for classification. If we classify real non-stationary signals, it is important to have informative classification features. Among such features are wavelet coefficients.

3. Harmonic Wavelets

Wavelet transform uses wavelets as the basis functions. An arbitrary function can be obtained from one function ("mother" wavelet) by using translations and dilations in the time domain. The wavelet transform is commonly used for analyzing non-stationary (seismic, biological, hydroacoustic etc.) signals, usually together with various spectral analysis algorithms [16,17].

Consider the basis of harmonic wavelets whose spectra are rectangular in the given frequency band [15,16]. Harmonic wavelets are usually represented in the frequency domain. Wavelet-function (mother wavelet) can be written as:

$$\Psi(\omega) = \begin{cases} \frac{1}{2\pi}, & 2\pi \leq \omega < 4\pi \\ 0, & \omega < 2\pi, \omega \geq 4\pi \end{cases} \Leftrightarrow \psi(x) = \int_{-\infty}^{\infty} \Psi(\omega) e^{i\omega x} d\omega = \frac{e^{i4\pi x} - e^{i2\pi x}}{i2\pi x} \qquad (1)$$

There are some techniques that allow us to decompose input signals using different basic functions: wavelets, sine waves, damped sine waves, polynomials, etc. These functions form the atom dictionary (basis functions) and each function is localized in the time and frequency domains. Often the dictionary of atoms is full (all types of functions are used) and redundant (the functions are not mutually independent). One of the main problems in these techniques is the selection of basic functions and dictionary optimization to acheive optimal decomposition levels [17]. Decomposition levels for wavelets can be defined as:

$$\Psi_{jk}(\omega) = \begin{cases} \frac{1}{2\pi} 2^{-j} e^{-\frac{i\omega k}{2^j}}, & 2\pi 2^j \leq \omega < 4\pi 2^j \\ 0, & \omega < 2\pi 2^j, \omega \geq 4\pi 2^j \end{cases}$$
$$\psi_{jk}(x) = \psi(2^j x - k) = \int_{-\infty}^{\infty} \Psi_{jk}(\omega) e^{i\omega x} d\omega = \frac{e^{i4\pi(2^j x - k)} - e^{i2\pi(2^j x - k)}}{i 2\pi(2^j x - k)} \quad (2)$$

where j is decomposition level and k is dilation.

Very often, wavelets are basis functions because of their useful properties [14] and the potential to process signals in the time-frequency domain. The Fourier transform of a scaling function can be written as:

$$\Phi(\omega) = \begin{cases} \frac{1}{2\pi}, & 0 \leq \omega < 2\pi \\ 0, & \omega < 0, \omega \geq 2\pi \end{cases} \Leftrightarrow \phi(x) = \int_{-\infty}^{\infty} \Phi(\omega) e^{i\omega x} d\omega = \frac{e^{i2\pi x} - 1}{i2\pi x} \quad (3)$$

We can formulate the following properties of harmonic wavelets, which relate them with other classes of wavelets:

- Harmonic wavelets have compact support in the frequency domain, which can be used for localizing signal features.
- There are fast algorithms based on the fast Fourier transform (FFT) for computing wavelet coefficients and reconstructing signals in the time domain.

The drawback of harmonic wavelets is their weak localization properties in the time domain in comparison with other types of wavelets. The spectrum in the form of a rectangular wave leads to decay in the time domain as $1/x$, which is not sufficient for extracting short-term singularities in a signal in the time domain.

Wavelet Transform in the Basis of Harmonic Wavelets

Detailed coefficients $a_{jk}, \widetilde{a}_{jk}$ and approximation coefficients $a_{\phi k}, \widetilde{a}_{\phi k}$:

$$\begin{aligned} a_{jk} &= 2^j \int_{-\infty}^{\infty} f(x) \overline{\psi}(2^j x - k) dx & \widetilde{a}_{jk} &= 2^j \int_{-\infty}^{\infty} f(x) \psi(2^j x - k) dx \\ a_{\phi k} &= 2^j \int_{-\infty}^{\infty} f(x) \overline{\phi}(x - k) dx & \widetilde{a}_{\phi k} &= 2^j \int_{-\infty}^{\infty} f(x) \phi(x - k) dx \end{aligned} \quad (4)$$

where j is the decomposition level; k is the dilation.

If $f(x)$ is a real-valued function, then: $\widetilde{a}_{jk} = \overline{a}_{jk}, \widetilde{a}_{\phi k} = \overline{a}_{\phi k}$.

Wavelet decomposition [14]:

$$f(x) = \sum_{j=-\infty}^{\infty} \sum_{k=-\infty}^{\infty} a_{jk} \psi(2^j x - k) = \sum_{k=-\infty}^{\infty} a_{\phi k} \phi(x - k) + \sum_{j=0}^{\infty} \sum_{k=-\infty}^{\infty} a_{jk} \psi(2^j x - k) \quad (5)$$

Wavelet decomposition using harmonic wavelets [18]:

$$\begin{aligned} f(x) &= \sum_{j=-\infty}^{\infty} \sum_{k=-\infty}^{\infty} \left[a_{jk} \psi(2^j x - k) = \widetilde{a}_{jk} \overline{\psi}(2^j x - k) \right] \\ &= \sum_{k=-\infty}^{\infty} \left[a_{\phi k} \phi(x - k) = \widetilde{a}_{\phi k} \overline{\phi}(x - k) \right] + \sum_{j=0}^{\infty} \sum_{k=-\infty}^{\infty} \left[a_{jk} \psi(2^j x - k) + \widetilde{a}_{jk} \overline{\psi}(2^j x - k) \right] \\ a_{jk} &= 2^j \int_{-\infty}^{\infty} f(x) \overline{\psi}(2^j x - k) dx \end{aligned} \quad (6)$$

Calculations with the last two formulae are inefficient.

Fast decomposition can be implemented in the following way:

$$a_{jk} = 2^j \int_{-\infty}^{\infty} F(\omega) \frac{1}{2\pi} 2^{-j} e^{\frac{i\omega k}{2^j}} d\omega = \frac{1}{2\pi} \int_{2\pi 2^j}^{4\pi 2^j} F(\omega) e^{\frac{i\omega k}{2^j}} d\omega \approx \int_{2\pi 2^j}^{4\pi 2^j} F(\omega) e^{\frac{i\omega k}{2^j}} d\omega \quad (7)$$

The substitution is of the following form:

$$n = 2^j + s$$
$$F_{2^j+s} = 2\pi F\big[\omega = 2\pi(2^j + s)\big] \quad (8)$$

We can show that:

$$a_{jk} = \sum_{s=0}^{2^j-1} F_{2^j+s} e^{\frac{i2\pi sk}{2^j}} \quad k = 0 \ldots 2^j - 1; \, j = 0 \ldots n - 1. \quad (9)$$

$$\widetilde{a}_{jk} = \sum_{s=0}^{2^j-1} F_{N-(2^j+s)} e^{\frac{i2\pi sk}{2^j}} \quad k = 0 \ldots 2^j - 1; \, j = 0 \ldots n - 1. \quad (10)$$

$$\widetilde{a}_{jk} = \bar{a}_{jk}$$

Thus, the algorithm for computing wavelet coefficients of the octave harmonic wavelet transform [19] of a continuous-time function $f(x)$ can be written in the following way:

1. The original function $f(x)$ is represented by discrete-time samples: $f(n)$, $n = 0 \ldots N-1$, where N is of degree 2 (if necessary, we use zero-padding).
2. We calculate the discrete Fourier transform using the fast Fourier transform to obtain a set of complex numbers $f(n)$, $n = 0 \ldots N-1$—Fourier coefficients (DFT coefficients).
3. Octave blocks F are processed using the discrete Fourier transform (DFT) to obtain coefficients: $a_{jk} = a_{2^j+k}$. The calculation results for the coefficients are given in Table 1.

Table 1. Distribution of wavelet coefficients among decomposition levels.

Number of Decomposition Level j	Wavelet Coefficients	Number of Wavelet Coefficients
−1	$a_0 = F_0$	1
0	$a_1 = F_1$	1
1	a_2, a_3	2
2	a_4, a_5, a_6, a_7	4
3	$a_8 \ldots a_{15}$	8
...	...	...
j	$a_{2^j} \ldots a_{2^{j+1}-1}$	2^j
...		...
$n-2$	$a_{N/4} \ldots a_{N/2-1}$	2^{n-2}
$n-1$	$a_{N/2}$	1

Further, consider two approaches to classifying bio-acoustic signals. We have used real hydroacoustic signals of whales from the database [20].

4. Classification Using the kNN-Algorithm

The classification was based on 14,822 records of whales of two types: 'killer' (4673 records) and 'pilot' (10,149 records). Data for processing was taken from [20]. Research has been conducted for the following signal-to-noise ratios (SNR): 100, 3, 0 and −3 dB. Training of the classifier was based on 85% of records of each class, and testing was based on 15% of records of each class. The following attributes have been used for comparison: the harmonic wavelet transform (HWT) coefficients, the short-time Fourier transform (STFT) coefficients and the discrete Fourier transform (DFT) coefficients.

All records had different numbers of samples (8064–900,771) and different sampling rates. To perform classification, we had to change the lengths of the records so that they equaled 2. To reduce the feature space dimension, we employed the approach based on modulo N reduction [21]. Such an approach allows us to reduce the data dimension when calculating N-point DFT if $N < L$ (L is signal length). The final signal matrix size ($N = 4096$) was 14,822 × 4096.

To reduce the feature space dimension, we also used coefficients of symmetry for the harmonic wavelet transform and the DFT: we used 50% coefficients (matrix: 14,822 × 2048). In the case of using a short-time Fourier transform (Hamming window of the size 256, overlap 50%), the final signal matrix size was 14,822 × 3999.

Below we can see the classification results (Tables 2–13, Figure 1) using the kNN-algorithm [22] for different features and different SNR values.

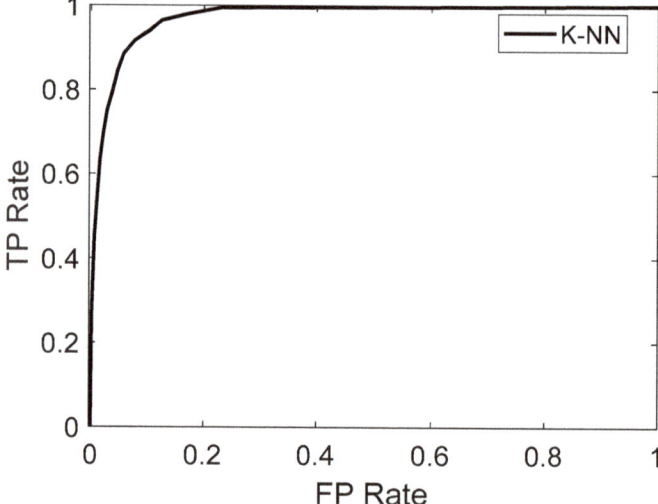

Figure 1. ROC curve of the classification: HWT, SNR = 100 dB.

The classification problem is to attribute vectors to different classes. We have two classes: positive and negative. In this case, we can have four different situations at the output of a classifier:

- If the classification result is positive, and the true value is positive as well, we have a true-positive value—TP.
- If the classification result is positive, but the true value is negative, we have false-positive value—FP.
- If the classification result is negative, and the true value is negative as well, we have a true-negative value—TN.
- If the classification result is negative, but the true value is positive, we have a false-negative value—FN.

We have calculated the following classification quality metrics: precision, recall and accuracy.

$$Precision = \frac{TP}{TP + FP}; \; Recall = \frac{TP}{TP + FN}; \; Accuracy = \frac{TP + TN}{TP + TN + FP + FN} \qquad (11)$$

Tables 14–16 contain precision, recall and accuracy for different classification features and different signal-to-noise ratios. Additionally, we can find the average final efficiency score characterizing the use of different classification features.

Table 2. Classification results: HWT, SNR = 100 dB.

Feature: HWT SNR = 100 dB	Killer	Pilot	Totals
Killer	TP = 590	FN = 110	700
Pilot	FP = 85	TN = 1437	1522
Totals	675	1547	2222

Table 3. Classification results: STFT, SNR = 100 dB.

Feature: STFT SNR = 100 dB	Killer	Pilot	Totals
Killer	TP = 591	FN = 109	700
Pilot	FP = 107	TN = 1415	1522
Totals	698	1524	2222

Table 4. Classification results: DFT, SNR = 100 dB.

Feature: DFT SNR = 100 dB	Killer	Pilot	Totals
Killer	TP = 592	FN = 108	700
Pilot	FP = 93	TN = 1429	1522
Totals	685	1537	2222
Totals	813	1409	2222

Table 5. Classification results: HWT, SNR = 3 dB.

Feature: HWT SNR = 3 dB	Killer	Pilot	Totals
Killer	TP = 642	FN = 58	700
Pilot	FP = 171	TN = 1351	1522

Table 6. Classification results: STFT, SNR = 3 dB.

Feature: STFT SNR = 3 dB	Killer	Pilot	Totals
Killer	TP = 642	FN = 58	700
Pilot	FP = 238	TN = 1284	1522
Totals	880	1342	2222

Table 7. Classification results: DFT, SNR = 3 dB.

Feature: DFT SNR = 3 dB	Killer	Pilot	Totals
Killer	TP = 535	FN = 165	700
Pilot	FP = 112	TN = 1410	1522
Totals	647	1575	2222

Table 8. Classification results: HWT, SNR = 0 dB.

Feature: HWT SNR = 0 dB	Killer	Pilot	Totals
Killer	TP = 669	FN = 31	700
Pilot	FP = 228	TN = 1294	1522
Totals	897	1325	2222

Table 9. Classification results: STFT, SNR = 0 dB.

Feature: STFT SNR = 0 dB	Killer	Pilot	Totals
Killer	TP = 646	FN = 54	700
Pilot	FP = 297	TN = 1225	1522
Totals	943	1279	2222

Table 10. Classification results: DFT, SNR = 0 dB.

Feature: DFT SNR = 0 dB	Killer	Pilot	Totals
Killer	TP = 439	FN = 261	700
Pilot	FP = 145	TN = 1377	1522
Totals	584	1638	2222

Table 11. Classification results: HWT, SNR = −3 dB.

Feature: HWT SNR = −3 dB	Killer	Pilot	Totals
Killer	TP = 674	FN = 26	700
Pilot	FP = 333	TN = 1189	1522
Totals	1007	1215	2222

Table 12. Classification results: STFT, SNR = −3 dB.

Feature: STFT SNR = −3 dB	Killer	Pilot	Totals
Killer	TP = 617	FN = 83	700
Pilot	FP = 336	TN = 1186	1522
Totals	953	1269	2222

Table 13. Classification results: DFT, SNR = −3 dB.

Feature: DFT SNR = −3 dB	Killer	Pilot	Totals
Killer	TP = 294	FN = 406	700
Pilot	FP = 144	TN = 1378	1522
Totals	438	1784	2222

Table 14. Classification results: HWT.

HWT	SNR = 100 dB		SNR = 3 dB		SNR = 0 dB		SNR = −3 dB	
Precision	0.8740	I *	0.7897	II *	0.7458	II *	0.6693	II *
Recall	0.8429	III *	0.9171	I *	0.9557	I *	0.9629	I *
Accuracy	0.9122	I *	0.8969	I *	0.8834	I *	0.8384	I *
Averaged score for three metrics	I		I		I		I	
Final score				I				

* score of a particular metric for each SNR. The "averaged score for three metrics" means that we estimated the average score for three metrics with the same SNR. Then, the final score for each feature (HWT, STFT, DFT) with different SNRs was chosen. We can see that using HWT as features gives the best result.

Table 15. Classification results: STFT.

STFT	SNR = 100 dB		SNR = 3 dB		SNR = 0 dB		SNR = −3 dB	
Precision	0.8467	III *	0.7295	III *	0.6850	III *	0.6474	III *
Recall	0.8443	II *	0.9171	I *	0.9229	II *	0.8814	II *
Accuracy	0.9028	III *	0.8668	III *	0.8420	II *	0.8114	II *
Averaged score for three metrics	III		III		II–III		II–III	
Final score					III			

* score of a particular metric for each SNR. The "averaged score for three metrics" means that we estimated the average score for three metrics with the same SNR. Then, the final score for each feature (HWT, STFT, DFT) with different SNRs was chosen. We can see that using HWT as features gives the best result.

Table 16. Classification results: DFT.

DFT	SNR = 100 dB		SNR = 3 dB		SNR = 0 dB		SNR = −3 dB	
Precision	0.8642	II *	0.8269	I *	0.7517	I *	0.6712	I *
Recall	0.8457	I *	0.7643	II *	0.6271	III *	0.4200	III *
Accuracy	0.9095	II *	0.8753	II *	0.8173	III *	0.7525	III *
Averaged score for three metrics	II		II		II–III		II–III	
Final score					II			

* score of a particular metric for each SNR. The "averaged score for three metrics" means that we estimated the average score for three metrics with the same SNR. Then, the final score for each feature (HWT, STFT, DFT) with different SNRs was chosen. We can see that using HWT as features gives the best result.

5. Classification Using a Deep Neural Network

The classification was based on 14,822 records of whales of two types: 'killer' (4673 records) and 'pilot' (10,149 records). Data for processing were taken from [20], containing sound recordings of 26 whales of two types: killer whale (15 individuals) and pilot whale (11 individuals).

In [23], for this dataset, two classifiers were constructed based on the kNN-algorithm. In the first case, the sounds were classified into a grind or killer whale sounds. For training, 800 whale sounds of each class were used; for testing, 400 of each were used. A classification accuracy of 92% was obtained. In the second experiment, 18 whales were separated from each other. For training, they took 80 records; for testing, they took 20. The classification accuracy was 51%.

In this work, records less than 960 ms long were removed from the dataset. After that, 14,810 records with an average duration of 4 s remained: 10,149 records of the grind and 4661 records of killer whales.

The classifier for both tasks was based on the VGGish model [24], which is a modified deep neural network VGG [25] pre-trained on the YouTube-8M dataset [26]. Cross entropy was used as a loss function. The audio files have been pre-processed in accordance with the procedure presented in [24]. Each record is divided into non-overlapping 960 ms frames, and each frame inherits the label of its parent video. Then log-mel spectrogram patches of 96 × 64 bins are then calculated for each frame. These form the set of inputs to the classifier. The output for the entire audio recording was carried out according to the maximum likelihood for all classes for each segment. As features, the output of the penultimate layer of dimension 128 was taken. More details can be found in the paper [23].

5.1. Experiment 1—Classification by Type

For the first task, we divided the dataset into training and test data in the proportion 85:15; in the training and the test sample there are no sounds from the same whales. The killer whale was designated 0, and the pilot whale was designated 1. Statistics on the training set: 8486–1, 3995–0. Statistics on the test set: 1663–1, 666–0.

The following results were obtained. On the training set, the confusion matrix was:

$$\begin{pmatrix} 3994 & 1 \\ 27 & 8459 \end{pmatrix}$$

1—FP, 27—FN.

On the test set, the confusion matrix was:

$$\begin{pmatrix} 633 & 33 \\ 86 & 1577 \end{pmatrix}$$

33—FP, 86—FN

Recall = 0.95, precision = 0.98 or accuracy = 0.95, AUC = 0.99.
Figure 2 shows the ROC curve for the test set.

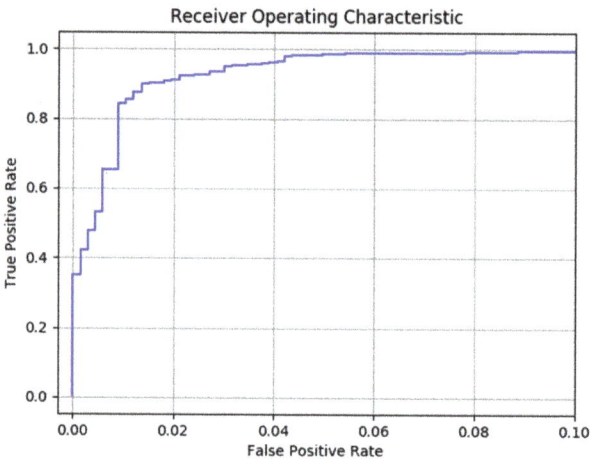

Figure 2. ROC curve for the test set.

5.2. Experiment 2—Classification by Individual

The data was divided into training and test sets in the ratio of 85:15, maintaining the proportions of the classes. As Figure 3 shows, the classes are very unbalanced. Thus, in the training test, for classes 26 and 5, 15 and 12 files are available. For class 20, 3684 files are available (see Figure 3).

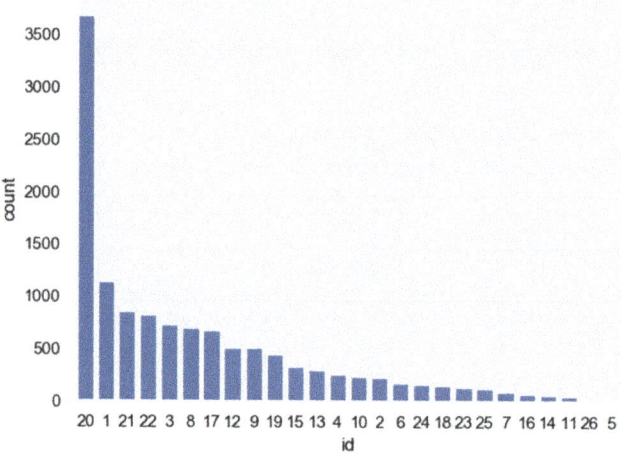

Figure 3. The number of files available for each class (individual). For training, from each class we took 900 files (augmented).

The confusion matrix for the training set is given in Figure 4.

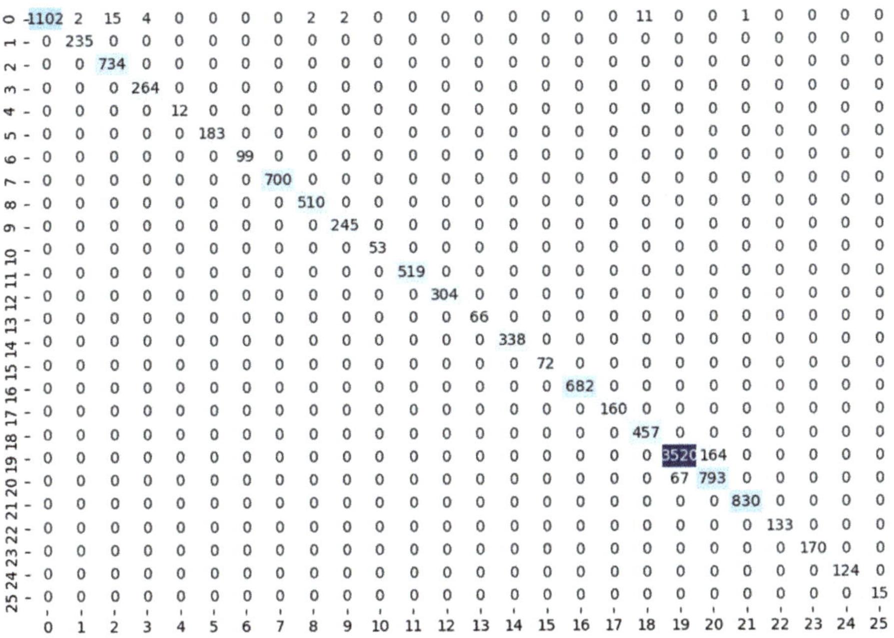

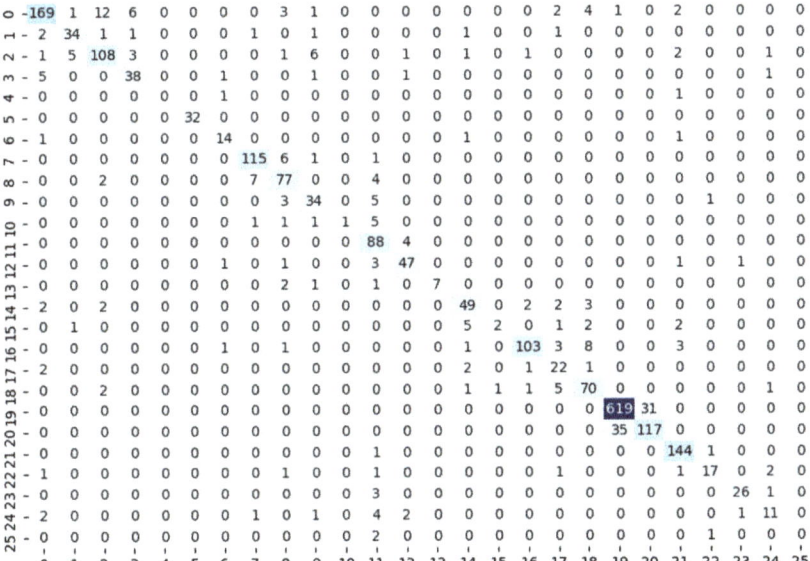

Figure 4. Confusion matrix for the training set.

The confusion matrix for the test set is presented in Figure 5.

Figure 5. Confusion matrix for the test set.

The accuracy of the classification of individuals in percent on a test sample is presented in Figure 6. Blue lines indicate the true-positive value, orange lines indicate false-positive value.

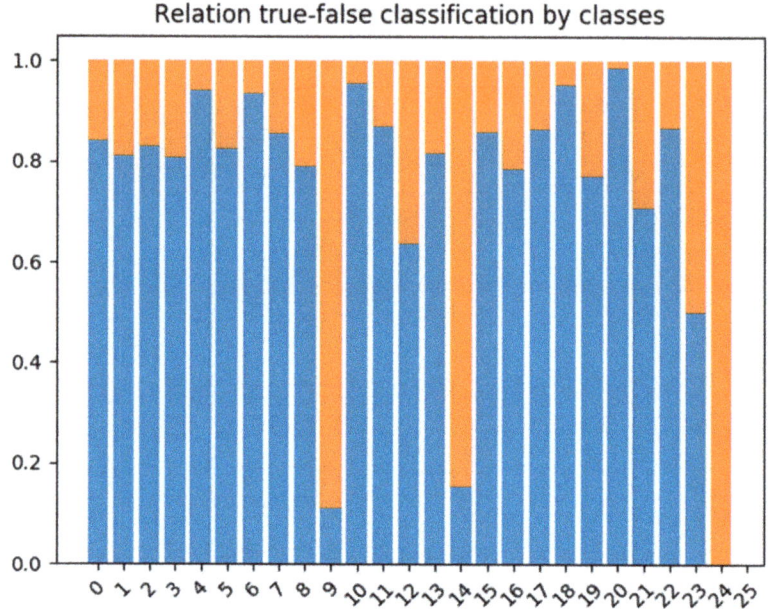

Figure 6. Classification accuracy for 26 whales.

As can be seen, the 25th (whale ID 26) class never predicts. Only for the 9th (whale ID 10), 14th (whale ID 15), 24th (whale ID 25) classes was the classification accuracy below 60%; for all the others it was higher. For some classes, classification accuracy is higher than 95%.

6. Discussion

Classification of whale sounds is a challenging problem that has been studied for a long time. Despite great achievements in feature engineering, signal processing and machine learning techniques, there still remain some major problems to be solved. In this paper, we used harmonic wavelets and deep neural networks. The results of the classification of whale types and individuals by means of deep neural networks are better than in previous works [23] with this dataset, but accuracy in the classification of types using harmonic wavelets as features and in the classification of individuals using deep neural networks should be increased. In further studies, we will use a Hilbert–Huang transform [27] and adaptive signal processing algorithms [28] to generate features.

For improvement of individual classification, two approaches can be suggested. The first combines data augmentation with other architectures of the neural network, but this will lead to large computational costs. The second approach is to use technology for simple and non-iterative improvements of multilayer and deep learning neural networks and artificial intelligence systems, which was proposed some years ago [29,30]. Our further research in the classification of hydroacoustic signals will be related to these two approaches. We also intend to test these approaches by adding noises at different SNRs, as we have done for harmonic wavelets.

7. Conclusions

In our paper, we considered the harmonic wavelet transform and its application to classifying hydroacoustic signals from whales of two types. We have provided a detailed representation of the mathematical tools, including fast computation of the harmonic wavelet transform coefficients. Classification results analysis allows us to draw conclusions about the reasonability of using harmonic wavelets when analyzing complex data. We have established that the smallest classification error is provided by the k-NN algorithm based on the harmonic wavelet transform coefficients.

The analysis (Table 17 Figures 5 and 6) illustrates the superiority of using a neural network for the Whale FM Project dataset in comparison with known work [23] and a kNN-classifier for the classification problem [31]. However, it is worth noting that the implementation of a neural network of such a complicated structure requires significant computational resources.

Table 17. Analysis of different approaches to bioacoustic signal classification of whale types (pilot and killer).

	Recall	Precision
Neural network	0.95	0.98
kNN-algorithm Feature: HWT SNR = 100 dB	0.84	0.87

Classification of 26 individual whales from the Whale FM Project dataset was proposed, and better results in comparison with previous works were achieved [23].

The proposed approach can be used in the study of the fauna of the oceans by research institutes, environmental organizations, and enterprises producing equipment for sonar monitoring. In addition, the study showed that the same methods can be used for speech processing and classification of underwater bioacoustic signals, which will subsequently allow the creation of effective medical devices based on these methods.

Author Contributions: Conceptualization, D.K.; data curation, S.R.; formal analysis, S.R. and D.B.; investigation, A.V. and S.R.; methodology, V.A. and D.B.; project administration, D.K.; resources, D.K. and V.A.; software, A.V. and V.A.; supervision, D.K.; validation, A.V., S.R. and D.B.; Visualization, A.V.; writing—original draft, D.K. and D.B.; writing—review & editing, V.A. and D.B. All authors have read and agreed to the published version of the manuscript.

Funding: The research and the present paper are supported by the Russian Science Foundation (Project NO. 17-71-20077).

Conflicts of Interest: The authors declare no conflict of interest.

References

1. Orcinus Orca (Killer Whale). Available online: https://www.iucnredlist.org/species/15421/50368125 (accessed on 16 March 2020).
2. Shiu, Y.; Palmer, K.J.; Roch, M.A.; Fleishman, E.; Liu, X.; Nosal, E.-M.; Helble, T.; Cholewaik, D.; Gillespie, D.; Klinck, H. Deep neural networks for automated detection of marine mammal species. *Sci. Rep.* **2020**, *10*, 607. [CrossRef] [PubMed]
3. Dréo, R.; Bouffaut, L.; Leroy, E.; Barruol, G.; Samaran, F. Baleen whale distribution and seasonal occurrence revealed by an ocean bottom seismometer network in the Western Indian Ocean. *Deep Sea Res. Part II Top. Stud. Oceanogr.* **2019**, *161*, 132–144. [CrossRef]
4. Bouffaut, L.; Madhusudhana, S.; Labat, V.; Boudraa, A.; Klinck, H. Automated blue whale song transcription across variable acoustic contexts. In Proceedings of the OCEANS 2019, Marseille, France, 17–20 June 2019; pp. 1–6.
5. Bouffaut, L.; Dréo, R.; Labat, V.; Boudraa, A.-O.; Barruol, G. Passive stochastic matched filter for antarctic blue whale call detection. *J. Acoust. Soc. Am.* **2018**, *144*, 955–965. [CrossRef] [PubMed]
6. Bahoura, M.; Simard, Y. Blue whale calls classification using short-time Fourier and wavelet packet transforms and artificial neural network. *Digit. Signal Process.* **2010**, *20*, 1256–1263. [CrossRef]
7. Zhong, M.; Castellote, M.; Dodhia, R.; Ferres, J.L.; Keogh, M.; Brewer, A. Beluga whale acoustic signal classification using deep learning neural network models. *J. Acoust. Soc. Am.* **2020**, *147*, 1834–1841. [CrossRef] [PubMed]

8. Zhang, L.; Wang, D.; Bao, C.; Wang, Y.; Xu, K. Large-Scale Whale-Call Classification by Transfer Learning on Multi-Scale Waveforms and Time-Frequency Features. *Appl. Sci.* **2019**, *9*, 1020. [CrossRef]
9. Smirnov, E. North Atlantic Right Whale Call Detection with Convolutional Neural Networks. In Proceedings of the ICML 2013 Workshop on Machine Learning for Bioacoustics, Atlanta, GA, USA, 16–21 June 2013.
10. Dorian, C.; Lefort, R.; Bonnel, J.; Zarader, J.L.; Adam, O. Bi-Class Classification of Humpback Whale Sound Units against Complex Background Noise with Deep Convolution Neural Network. Available online: https://arxiv.org/abs/1703.10887 (accessed on 12 March 2019).
11. Hodges, R.P. *Underwater Acoustics: Analysis, Design, and Performance of Sonar*; John Wiley & Sons: London, UK, 2010.
12. Kaplun, D.; Klionskiy, D.; Voznesenskiy, A.; Gulvanskiy, V. Digital filter bank implementation in hydroacoustic monitoring tasks. *PRZ Elektrotechniczn* **2015**, *91*, 47–50. [CrossRef]
13. Milne, P.H. *Underwater Acoustic Positioning Systems*; Gulf Publishing Co.: Houston, TX, USA, 1983.
14. Mallat, S. *A Wavelet Tour of Signal Processing*, 3rd ed.; Academic: New York, NY, USA, 2008.
15. Klionskiy, D.M.; Kaplun, D.I.; Gulvanskiy, V.V.; Bogaevskiy, D.V.; Romanov, S.A.; Kalincev, S.V. Application of harmonic wavelets to processing oscillating hydroacoustic signals. In Proceedings of the 2017 Progress in Electromagnetics Research Symposium—Fall (PIERS—FALL), Singapore, 19–22 November 2017; pp. 2528–2533.
16. Newland, D.E. Harmonic wavelet analysis. Series A. *Proc. R. Soc. Lond.* **1993**, *443*, 203–225.
17. Kaplun, D.; Voznesenskiy, A.; Romanov, S.; Nepomuceno, E.; Butusov, D. Optimal Estimation of Wavelet Decomposition Level for a Matching Pursuit Algorithm. *Entropy* **2019**, *21*, 843. [CrossRef]
18. Newland, D.E. *An Introduction to Random Vibrations, Spectral & Wavelet Analysis*, 3rd ed.; John Wiley & Sons: New York, NY, USA, 1993.
19. Newland, D.E. Harmonic wavelets in vibrations and acoustics. *Philos. Trans. R. Soc. A* **1999**, *357*, 2607–2625. [CrossRef]
20. Whale FM Project. Available online: https://whale.fm (accessed on 3 April 2020).
21. Orfanidis, S.J. *Introduction to Signal Processing*; Prentice Hall: Upper Saddle River, NJ, USA, 1996.
22. Mitchell, T. *Machine Learning*; McGraw-Hill: New York, NY, USA, 1997.
23. Shamir, L.; Yerby, C.; Simpson, R.; von Benda-Beckmann, A.M.; Tyack, P.; Samarra, F.; Miller, P.; Wallin, J. Classification of large acoustic datasets using machine learning and crowdsourcing: Application to whale calls. *J. Acoust. Soc. Am.* **2014**, *135*, 953–962. [CrossRef] [PubMed]
24. Hershey, S.; Chaudhuri, S.; Ellis, D.P.W.; Gemmeke, J.F.; Jansen, A.; Moore, C.; Plakal, M.; Platt, D.; Saurous, R.A.; Seybold, B.; et al. CNN architectures for largescale audio classification. In Proceedings of the International Conference on Acoustics, Speech and Signal Processing (ICASSP), New Orleans, LA, USA, 5–9 March 2017.
25. Simonyan, K.; Zisserman, A. Very Deep Convolutional Networks for Large-Scale Image Recognition. *arXiv* **2015**, arXiv:1409.1556v6.
26. A Large and Diverse Labeled Video Dataset for Video Understanding Research. Available online: https://research.google.com/youtube8m/ (accessed on 3 April 2020).
27. Huang, N.E.; Shen, S.S.P. *Hilbert-Huang Transform and Its Applications*; World Scientific: Singapore, 2005; 350p.
28. Voznesensky, A.; Kaplun, D. Adaptive Signal Processing Algorithms Based on EMD and ITD. *IEEE Access* **2019**, *7*, 171313–171321. [CrossRef]
29. Tyukin, I.Y.; Gorban, A.N.; Prokhorov, D.V.; Green, S. Efficiency of Shallow Cascades for Improving Deep Learning AI Systems. In Proceedings of the 2018 International Joint Conference on Neural Networks (IJCNN), Rio de Janeiro, Brazil, 8–13 July 2018; pp. 1–8.
30. Gorban, A.N.; Burton, R.; Romanenko, I.; Tyukin, I.Y. One-trial correction of legacy AI systems and stochastic separation theorems. *Inf. Sci.* **2019**, *484*, 237–254. [CrossRef]
31. Marsland, S. *Machine Learning an Algorithmic Perspective*, 2nd ed.; CRC Press: Boca Raton, FL, USA, 2016.

© 2020 by the authors. Licensee MDPI, Basel, Switzerland. This article is an open access article distributed under the terms and conditions of the Creative Commons Attribution (CC BY) license (http://creativecommons.org/licenses/by/4.0/).

Article

Quantification of the Feedback Regulation by Digital Signal Analysis Methods: Application to Blood Pressure Control Efficacy

Nikita S. Pyko [1], Svetlana A. Pyko [1,2], Oleg A. Markelov [1,*], Oleg V. Mamontov [2,3] and Mikhail I. Bogachev [1,*]

[1] Radio Systems Department & Biomedical Engineering Research Centre, Saint Petersburg Electrotechnical University, 5 Professor Popov Street, 197376 Saint Petersburg, Russia; goststalker13@gmail.com (N.S.P.); svet.pyko@gmail.com (S.A.P.)
[2] Department for Cardiovascular Physiology, Almazov National Medical Research Centre, 2 Akkuratova Street, 197341 Saint Petersburg, Russia; mamontoffoleg@gmail.com
[3] Faculty Therapy Department, Pavlov First State Medical University, 6/8 Leo Tolstoy Street, 197022 Saint Petersburg, Russia
* Correspondence: oamarkelov@etu.ru (O.A.M.); rogex@yandex.ru (M.I.B.)

Received: 7 November 2019; Accepted: 18 December 2019; Published: 26 December 2019

Featured Application: Analysis of blood pressure and heart rate mutual synchronization provides complementary information about the physiological mechanisms and efficacy of feedback control that keeps blood pressure levels within the physiologically desirable range.

Abstract: Six different metrics of mutual coupling of simultaneously registered signals representing blood pressure and pulse interval dynamics have been considered. Stress test responses represented by the reaction of the recorded signals to the external input by tilting the body into the upright position have been studied. Additionally, to the conventional metrics like the joint signal coherence Coher and the sensitivity of the pulse intervals response to the blood pressure changes baroreflex sensitivity (BRS), also alternative indicators like the synchronization coefficient Sync and the time delay stability estimate TDS representing the temporal fractions of the analyzed signal records exhibiting rather synchronous dynamics have been determined. In contrast to BRS, that characterizes the intensity of the pulse intervals response to the blood pressure changes during observed feedback responses, both Sync and TDS likely indicate how often such responses are being activated in the first place. The results indicate that in most cases BRS is typically reciprocal to both Sync and TDS suggesting that low intensity of the feedback responses characterized by low BRS is rather compensated by their more frequent activation indicated by higher Sync and TDS. The proposed additional indicators could be complementary for the differential diagnostics of blood pressure regulation efficacy and also lead to a deeper insight into the involved concomitant factors this way also aiming at the improvement of the mathematical models representing the underlying feedback control mechanisms.

Keywords: feedback regulation; digital signal analysis; control efficacy

1. Introduction

Blood pressure levels are being simultaneously affected by multiple internal and external factors that require continuous activity of several regulatory mechanisms to keep its levels within a certain homeostatic range. Among multiple mechanisms that are involved in the regulation of blood pressure, arterial baroreceptor reflex or simply, baroreflex, appears one of the key mechanisms that govern short-term feedback responses to various physical stresses such as exercise, adaptation to the changes in the body position, reactions to drugs, changes in the subject's mental conditions and so on. The increase of blood pressure is sensed by baroreceptors located in blood vessels that in turn invoke a response by the autonomous nervous system. The response is generally twofold, including the decrease of heart rate and the reduction of the vascular resistance, both leading to the consequent drop of blood pressure. Thus the efficacy of this feedback mechanism largely determines timely and adequate responses to the changes in blood pressure.

Over a long time quantitative assessment of the short-term blood pressure—heart rate feedback regulation efficacy has been limited to the analysis of the baroreflex sensitivity (BRS) defined as the measure of the relative change of the pulse interval (in ms) in response to the change in the systolic blood pressure (in mmHg), altogether measured in (ms/mmHg). Historically, BRS was first measured as the increase in the pulse interval to the pharmacologically induced increase in blood pressure that guaranteed the feedback mechanism activation for a certain time fragment, with the BRS being quantified by the linear regression coefficient of pulse intervals on blood pressure over this time fragment. In the last three decades, several methods to measure BRS from simultaneously recorded spontaneous fluctuations of both pulse intervals and blood pressure without applying any external stimuli have been suggested. The most common time-domain approach, often termed as the sequence method, simply focuses on finding time fragments where both pulse intervals and blood pressure either increase or decrease consecutively and monotonously over several heartbeat cycles, sometimes referred to as baroreflex sequences. Next for each time fragment a linear regression coefficient of pulse intervals on blood pressure is calculated, representing the local BRS estimate [1,2]. To improve the accuracy, averaging over several baroreflex sequences is usually performed, at the cost of lower temporal resolution. In contrast, spectral-based methods do not require selection of certain time fragments, instead focusing on the blood pressure—pulse intervals transfer function analysis in a certain frequency band where the coherence between them exceeds a certain threshold [3–5]. To overcome the common drawback of both methods, in particular their limited performance under non-stationary conditions such as stress tests, modified methods based on first differences analysis have been suggested [6]. Combined with the advances of non-invasive blood pressure measurement techniques, these methods made BRS one of the routinely measured parameters both in clinical and ambulatory settings [7].

BRS has been reported as a highly informative prognostic marker widely applicable in both ambulatory and clinical investigations. In particular, BRS appeared highly predictive of cardiac mortality in post-infarction patients with both reduced ejection fraction [8] as well as preserved left ventricular function [9] including those receiving β-blocker treatment [10] as well as in patients with life-threatening arrhythmias [11], see also [12]. BRS impairment has been also reported as an early indicator of autonomic dysfunction and autonomic failure [13]. In earlier studies, BRS has been shown to exhibit significant changes under exercise as well as postural and other physical stresses [14–17], while more recent data indicate that these effects are temporary and after a certain adaptation period the baroreflex exhibits a resetting around new absolute blood pressure and pulse interval values [18–20]. However, successful baroreflex resetting has notable exceptions with one such observed recently in diabetes patients with certain complications, particularly with obesity who exhibited significant baroreflex impairment [21].

While the BRS quantifies explicitly the response of the heart rate to the changes in blood pressure, it does not contain any information whether there was such a response to every significant variation of blood pressure. Time-domain methods simply disregard time intervals without significant changes of heart rate irrelevant to the fact whether there were blood pressure variations, while spectral methods provide characteristics that are averaged over a given frequency band within the entire analysis window. In turn, another alternative method [22] measures the average heart rate acceleration or deceleration, while not considering whether it occurred in response to significant blood pressure variations or not. Thus, neither of them can guarantee that all blood pressure variations were adequately responded. However, a timely response to the changes in blood pressure is essential for homeostasis, since missing or delayed responses lead to the increased blood pressure variability. In contrast, there is recent evidence that baroreflex activation therapy using implantable devices that stimulates carotid baroreceptors significantly improved the blood pressure control efficacy [23]. Accordingly, additionally to the measurement of the BRS itself, it is also important to quantify the activation of the feedback mechanism in response to blood pressure variations.

In this paper, we suggest a series of complementary indicators of the blood pressure—heart rate feedback regulation based on their mutual synchronization patterns. Investigation of the physiological signals' mutual synchronization is widely used in chronobiological studies appearing essential for a deeper understanding of the mechanisms related to the influence of various external factors such as geomagnetic field variations, solar cycles, jetlag or night shifts [24–26]. Recent examples include the synchronization analysis of heart rate and respiration during different sleep phases [27–29]. While in most cases relation between rhythms with certain quasi-periodic structure is studied, such as breathing cycles modulating heartbeat cycles, we go beyond that and, while following conceptually similar methodology, modify the synchronization analysis methodology to suite the blood pressure—pulse interval analysis that both exhibit rather stochastic behavior, as indicated below. Particular mutual synchronization metrics used here follow a recent study where their performance has been validated using simulated datasets with non-periodic structure and correlation patterns reminiscent to those typically observed in physiological signals [30].

2. Materials and Methods

2.1. Subjects and Clinical Investigation Protocol

All recordings were obtained at the Almazov National Medical Research Centre in accordance with the ethical standards presented in the Declaration of Helsinki. The study protocol was reviewed and approved by the Ethics Committee of the Almazov National Medical Research Centre (Ref. No. 110, approval date 12 June 2010) before the beginning of the study. All patients and volunteers provided their informed consent in written form prior to their participation in the study.

The study included 95 subjects subdivided into three groups:

- Group 1 included 33 patients aged 39.4 ± 17.5 years old with either classic or gradual orthostatic hypotension with orthostatic systolic blood pressure (SBP) reduction by at least 20 mmHg and/or by diastolic blood pressure reduction by at least 10 mmHg within three minutes of the orthostatic compared to the initial (supine) position. Functional tests performed prior to the tilt test indicated that most patients in this group demonstrated moderate autonomic dysfunction (see Table 1 for the functional autonomic tests results);
- Group 2 included 34 diabetes mellitus patients aged 52.7 ± 10.9 years old with autonomic neuropathy diagnosed by functional autonomic tests as indicated in [31,32] that were performed prior to the tilt test. Only patients with at least two (out of seven) reduced functional autonomic test responses (for further details, see the functional tests specification below) were considered. All diabetes patients received standard combination therapy of blood sugar reducing drugs, including glitazones, gliptins, metformin and sulfonylureas, while receiving no specific therapy of autonomic function within one month before the study;
- Group 3 contained of 28 healthy volunteers aged 41.2 ± 11.1 years old.

Table 1. Detailed clinical characteristics of the studied patients' groups.

Attribute/Parameter/Quantity	Group 1	Group 2
Age, years	39.4 ± 17.5	52.7 ± 10.9
Concomitant cardiovascular disease (arterial hypertension, coronary heart disease, heart failure), n (%)	21 (64)	32 (94)
Orthostatic decrease of systolic/diastolic blood pressure, mmHg	34 ± 13/17 ± 7	17 ± 14/2 ± 16
Reduced indicators of functional autonomic tests (out of seven), n	2.5 ± 1.1	5.0 ± 1.2
Diabetes duration, years	n/a	10.1 ± 3.9
Diabetes complications, n (%)	n/a	34 (100)

Prior to the tilt-test all patients and volunteers underwent standard functional autonomic tests. A comprehensive assessment of autonomic regulation of blood circulation included the following tests:

1. Valsalva maneuver (cardiac component);
2. Valsalva maneuver (vasomotor component);
3. Deep breathing test;
4. Hand-grip test;
5. Cold-stress induced vasoconstriction for evaluation reaction of the forearm vessels while cooling the skin of the upper chest for 2 min;
6. Spontaneous arterial baroreflex (ABR);
7. Spectral power of heart rate variability.

Next all 95 subjects and patients underwent a head-up tilt-test (table tilt 70°, duration of orthostatic position up to 30 min unless stopped earlier due to syncope response) [33], see also [34] that was performed under identical conditions between 10 am and 1 pm. The recording in the initial supine position and the initial fragment of the orthostatic phase recording (both around 10 min duration) were used in further analysis.

Hemodynamic parameters we measured continuously using the Finometer-Pro blood pressure monitor (Finapres Medical Systems, Enschede, The Netherlands) with parallel electrocardiogram (ECG) recording. The forearm blood flow was measured by venous occlusion plethysmography using Dohn air-filled cuff.

The overall study design is summarized in Figure 1.

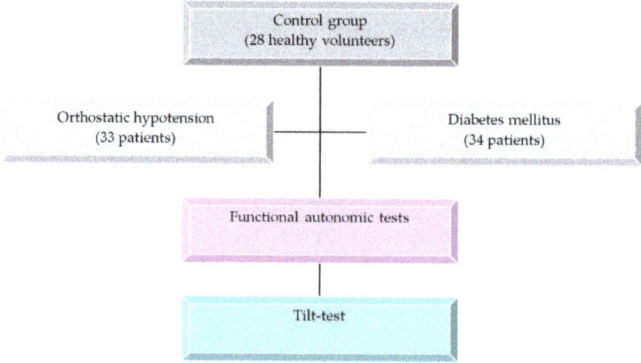

Figure 1. Clinical study design.

Detailed characteristic of the studied patients' groups are summarized in Table 1.

2.2. Data Acquisition and Preliminary Processing

Since the instantaneous phase calculation procedure by Hilbert transform employed in the signal mutual synchronization analysis is rather sensitive to random errors in measurements, an adaptive recursive filtering procedure has been applied to the original measurement sequences $\{s_i\}$, where s denotes either systolic blood pressure (SBP) or pulse intervals (PI) aiming at the elimination of the anomalous measurements. This procedure is based on the analysis of the first differences of the pulse intervals and systolic blood pressure values. The threshold value for the exclusion of outliers is based on the analysis of their empirical distribution functions for the processed data series. The elimination procedure consists of two consecutive steps that are repeated iteratively: (i) marking of the potential outliers as candidates for future elimination and (ii) removing marked outliers. In the first step, one calculates the first differences $s'_i = s_{i+1} - s_i$ from the initial dataset $\{s_i\}$ (that will be recalculated iteratively each time after a single outlier is eliminated). To mark the ith element of the dataset $\{s_i\}$ as a candidate for being an outlier ("OUT") three conditions should be met simultaneously:

1. The absolute values of the first differences $|s'_{i-1}| = |s_i - s_{i-1}|$ and $|s'_i| = |s_{i+1} - s_i|$ that are adjacent to the measured data $\{s_i\}$ are both either greater or equal than the standard deviation $\sigma\{s_i\}$ of the first differences.
2. The two consecutive first differences s'_{i-1} and s'_i have different signs.
3. The absolute values of the first differences normalized by its standard deviation $|s'_{i-1}|/\sigma\{s'_i\}$ and $|s'_i|/\sigma\{s'_i\}$ both exceed the critical value $\sigma_{\text{crit.}}(N)$ determined as a quantile of the Student's t-distribution for the given dataset size N.

Those elements for which all three conditions are met are then marked as "OUT". Out of the marked data $\{s_i\}$ the one with the largest normalized standard deviation is eliminated. After elimination of a single outlier the above procedure is repeated iteratively unless there are no values marked as outliers for the chosen elimination depth. For deeper details on the filtering algorithm, we refer to [35].

Since the sequence of pulse intervals was non-equidistant due to its inherent variability, we next used the cubic interpolation and resampling with the desired sampling frequency, 5 Hz in our case. Therefore, both analysed datasets were now represented by sequences equidistant in time and taken at the same time points.

2.3. BRS Estimation

To calculate BRS from blood pressure and pulse interval recordings during tilt tests, we followed a recently suggested methodology that is particularly suited for dealing with non-stationary data [6]. The first differences of SBP and PI values were taken and the BRS was estimated as a linear regression coefficient of ΔPI on ΔSBP in those quadrants where the signs of ΔPI and ΔSBP were identical. To disregard uncertain as well as anomalous variations beat-to-beat changes of less than 1 mmHg in SBP or less than 3 ms in PI and more than 20 mmHg in SBP or more than 100 ms in PI were ignored.

2.4. The Method of PI and BRS Phase Synchronization Measurement

To estimate the mutual synchronization behaviour of the two signals quantitatively we used the method based on the comparison of their phases [36]. Instantaneous phase values were determined by the Hilbert transform which is widely used in mathematics, physics and signal analysis. For the overall estimation algorithm design, see Figure 1.

The Hilbert transform produces complex function $\dot{s}(t)$ from the original real signal $s(t)$ (that stands for either SBP or PI) by adding the imaginary component $s_\perp(t)$, which is defined as:

$$s_\perp(t) = \frac{1}{\pi} \int_{-\infty}^{\infty} \frac{s(\tau)}{t - \tau} d\tau. \tag{1}$$

The resulting complex function is known as the complex analytical signal $\dot{s}(t) = s(t) + js_\perp(t)$.

The real and imaginary parts of the analytical signal allowed us to determine the envelope $S(t)$ as the absolute value of the analytic signal that characterized the laws governing its amplitude modulation, and the phase $\Phi(t)$ as the argument of the analytical signal that characterized the laws governing its angular modulation. Accordingly;

$$S(t) = \sqrt{s^2(t) + s_\perp^2(t)}, \; \Phi(t) = \arctg \frac{s_\perp(t)}{s(t)}. \tag{2}$$

Of note, the signal phase has a clear physical interpretation only in the case of harmonic or narrowband oscillations, while the above formalism was not restricted to these assumptions and thus allowed us to calculate phase values for arbitrary data sequences. Next, we determined the differences between the phases Φ_{PI}–Φ_{SBP}. Following [27,28] we next applied moving average filtering in a gliding window of size τ and calculated the standard deviation of the phase differences. Consecutive phase points where the standard deviation remained below a given threshold, equal to $2\pi/\delta$, were treated as belonging to synchronization episodes, once their duration exceeded T seconds. This procedure was applied to the entire record in a gliding window while counting episodes of synchronous behavior. As a result, a quantitative measure of the phase synchronization was the synchronization coefficient Sync defined as the percentage of the synchronous behaviour episodes duration within the total duration of the analysis window.

2.5. Adjustment of Synchronization Analysis Algorithm Parameters

For the initial adjustment of the methods and finding appropriate parameters of the synchronization analysis algorithm that fit to the typical blood pressure and heart rate variability characteristics, another set of 150 stationary records which were obtained independently from this study and already used in previously reported analysis [37] from subjects with various autonomic status under supine resting conditions (typical record duration around 10 min) have been used.

First optimization of synchronization algorithm parameters τ, T and δ was performed. The appropriate choice of these values depended on the specific experimental conditions and was not always universal for the given type of data. The parameters may be adjusted to optimize the sensitivity of the algorithm by avoiding the saturation at either very low or very high synchronization coefficients. The gliding window duration τ determined the number of phase points in the window. It was connected with the parameter δ used to calculate the threshold for the standard deviation of the instantaneous phase first differences since the standard deviation calculated for a finite data sample depended on the sample size.

To choose the appropriate window size τ, first the boundary conditions that specify its possible range were determined. One of the common hypotheses of the Mayer waves origin is their baroreflex loop based nature, suggesting that their period of about 10 s corresponds to the full feedback loop cycle [38]. Accordingly, any internally or externally induced changes in blood pressure were followed by characteristic regulatory oscillations with the Mayer waves period. Thus, choosing the gliding window size that was comparable or above this 10 s period eliminated these short-term regulatory oscillations by averaging. Alternatively, to ensure that the observed variations in both blood pressure and pulse intervals were proper measurements not caused by single faulty measurements and do not appear artifacts of preliminary filtering procedures, one has to guarantee that at least several actual measurements have been performed in each window. Taking into account typical heart rate values, a 3 s gliding window will typically result in having 3–4 pulse measurements under resting conditions and even more under stress conditions with an increased heart rate. Accordingly, one has to restrict with τ above 3 s, and preferably, not longer than half period of Mayer oscillations, that is 5 s. For a better temporal resolution of the analysis, choosing the lower bound of the appropriate range that was a 3 s gliding window seems plausible. Next the threshold for the standard deviation of the phase difference was chosen empirically as $2\pi/300$, $T = 0.4$ s, the latter adjusted empirically to avoid saturation and this way increase the dynamic range of the Sync index.

The graphic insets in Figure 2 exemplify the analysis results for a single tilt-test record. In addition to pre-processed data sequences and their Hilbert phases, the figure displays their first differences as well as the results of their standard deviation analysis in a gliding window (in the lower right panel). The dashed lines denote the $2\pi/\delta$ threshold for the standard deviation of the phase difference. The bold solid curve denotes $\Phi_{PI}-\Phi_{SBP}$ and curve fragments highlighted by red color denote the synchronization episodes where the standard deviation of the phase difference remains below the $2\pi/\delta$ threshold. Finally, the fraction of such episodes within the total record duration determines the synchronization coefficient Sync.

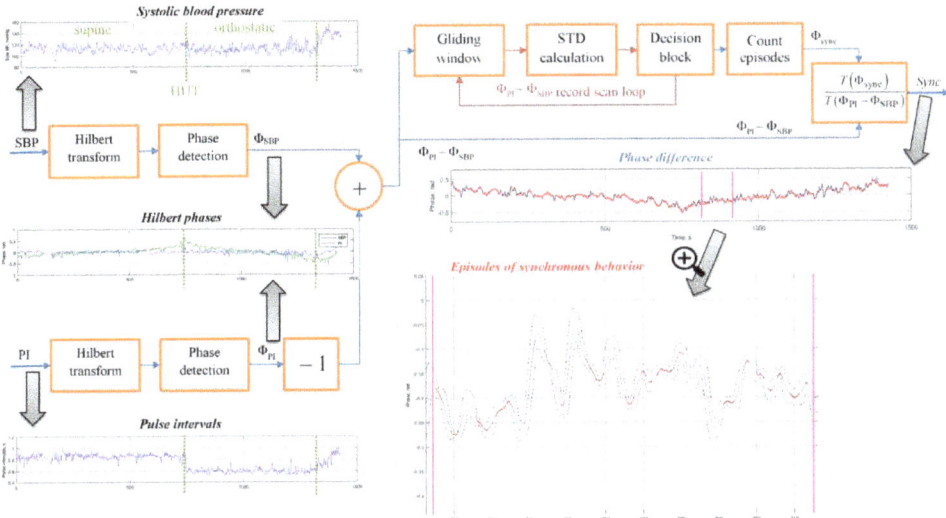

Figure 2. Systolic blood pressure vs. pulse intervals synchronization analysis algorithm design.

The figure shows that, while for the entire analysed fragment the synchronization coefficient Sync was somewhat around 50%, it exhibited considerable variations along the record. While the first part of the record was characterized by prolonged episodes of synchronous behaviour interrupted by few short-term asynchronous fragments, the second part of the fragment exhibited more frequent asynchronous behaviour episodes interrupted by rather few coupling patterns. While the particular reasons for each onset and breakdown of this coupling can hardly be determined, there was a clear discrepancy between the first and the second part of the record in terms of their synchronization patterns. Such changes in the synchronization behaviour of blood pressure and pulse intervals could have been triggered by some physical or mental stress like change in the body position, and so on. Accordingly, reactions to various stress patterns that were imposed during functional tests can be studied in terms of the changes in the blood pressure—the pulse intervals synchronization coefficient, Sync. This requires that the recordings being analysed during different test phases. For example, for the head-up tilt table testing the supine and the orthostatic test, phases could be analysed separately, this way allowing us to evaluate how the synchronization pattern changes in response to the orthostatic stress. An additional advantage of the proposed methodology is that the evaluation of the degree of phase synchronization between SBP and PI may be useful in the study of regulatory functions in the human body during various functional tests, since it does not require data stationarity.

While the BRS value characterizes the intensity of the heart rate reaction to the blood pressure changes during observed feedback responses, the Sync value likely indicates how often such responses are activated in the first place. Accordingly, low intensity of the reaction characterized by low BRS that should result in higher than normal blood pressure variability, theoretically, could be at least partially compensated by its more frequent activation characterized by higher Sync.

2.6. Alternative Mutual Information Metrics

While according to the results of a recent study [30], Sync appears the most sensitive of various mutual synchronization indices, it has its own drawbacks, including high sensitivity to any (including random) variations in the analyzed signals, that requires seeking for more robust alternatives that would respond more specifically only to significant variations, although typically at the cost of their lower sensitivity.

Among them, the third approach is the time delay stability estimate is based on the analysis of the relative shift of the maximum of the cross-correlation function of two studied series originally proposed in [28]. In this approach, the average delay in 50% overlapping windows of fixed length is calculated, and the time delay stability episode is determined once within at least five consecutive windows the shift of the maximum of the cross-correlation function remains below a given threshold. Like in the previous method, in order to estimate the time delay stability coefficient for an entire record, the TDS value is determined as the fraction of the time delay stability episodes in the total record duration. Similarly, the starting set of the parameters used in this study follows the results of a recent simulated data based investigation [30].

One more quantity utilized here is a certain combination of the previous two approaches and is based on the analysis of the correlation time of the phase differences of the studied data series. Like in the first method, the phase differences $K\Phi i(\tau)$ are determined using the Hilbert transform. Next the observational data series are divided into 50% overlapping time windows of duration T, and in each time window the correlation time of the phase differences is calculated as:

$$TAU_i = \frac{\int_0^T |K_{\Phi i}(\tau)| d\tau}{K_{\Phi i}(0)}. \qquad (3)$$

To obtain the overall statistics for a given data series, the averaged correlation time over all studied time windows is calculated. Once the window size is well above observed correlation times, this method becomes parameter-free.

The fifth approach utilizes the mean coherence of two data series [3]. Contrasting with the three above described methods that are all obtained in the time domain, the coherence function is obtained in the frequency domain and is calculated as:

$$C_{xy}(f) = \frac{|P_{xy}(f)|^2}{P_x(f)P_y(f)}. \qquad (4)$$

where $P_x(f)$ and $P_y(f)$ are the individual spectral densities, and $P_{xy}(f)$ is the cross-spectral density of the analyzed datasets.

The sixth method is based on the calculation of cross-conditional entropy of the two series $x(t)$ and $y(t)$ according to the approach described in [18]. First, the analyzed data series are normalized. Then in the series $y(t)$ the patterns including $L - 1$ samples are selected. The cross-conditional entropy is then defined as:

$$CE(L) = -\sum_{k=1}^{M} p(fx_k) \sum_{i=1}^{N} p(y(i)/fx_k) \times \log(p(y(i)/fx_k)), \qquad (5)$$

where fx is a data fragment of size $L - 1$ selected from the first data series $x(i)$; $p(fx)$ is the probability of the observation of fx within the series $y(t)$; is the number of corresponding fragments; $p(y(i)/fx)$ is the conditional probability of a particular sample $y(i)$ to be observed within a series $y(t)$ following the fx pattern.

The above index represents the amount of information carried by the sample $x(i)$ when the pattern fx is assigned. Thus, the coefficient CE depends on the pattern size L. It reaches zero when a sufficient number of samples of $y(t)$ carries the entire information of behavior of $x(t)$. It remains high and constant if the processes $x(t)$ and $y(t)$ are independent and yields intermediate values when knowledge of $y(t)$ allows for a limited prediction of the behavior of $x(t)$.

Figure 3 exemplifies the entire analysis procedure for a single tilt-test record. After preliminary data preparation both SBP and PI sequences are subjected to the Hilbert transform and phase detection as indicated above. The entire algorithm was implemented using the Matlab software package.

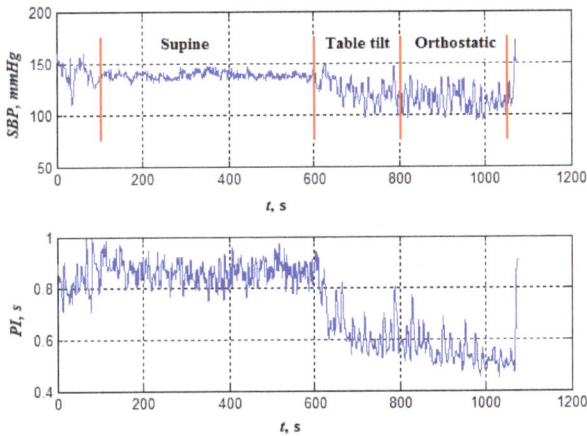

Figure 3. Typical dynamics of blood pressure and pulse intervals during a head-up tilt test, vertical red lines denote the test phases.

2.7. Statistical Analysis

Since our preliminary studies [35–37] indicated that the studied blood pressure—pulse intervals coupling metrics are not normally distributed, we used the methods of non-parametric statistics to process our results. To quantify the statistical significance of our results, we next applied the non-parametric Mann–Whitney U-test for independent samples [39]. To calculate the correlations between the synchronization coefficient, the standard deviations of pulse intervals and systolic blood pressure and the baroreflex sensitivity we used the non-parametric Spearman's correlation coefficients. All statistical analysis was performed using the IBM SPSS Statistics software package.

3. Results

Remarkably, while the orthostatic hypotension patients (Group 1) were characterized by a smaller number of reduced autonomic tests results (5.0 ± 1.2 vs. 2.5 ± 1.1) suggesting rather moderate autonomic dysfunction particularly less severe than that one in the diabetes patients (Group 2), they were characterized by considerably more pronounced blood pressure reduction during tilt test (34 ± 13 vs. 17 ± 14 for systolic and 17 ± 7 vs. 2 ± 16 for dyastolic blood pressure, respectively).

Analysis of both the entire tilt test records as well as each of its phases revealed that, as expected, the BRS differs significantly between healthy individuals and both patient groups, according to the non-parametric Mann–Whitney U-test ($p < 0.05$ for the first group with orthostatic hypotension and $p < 0.001$ for the second group of diabetes patients with autonomic neuropathy). Remarkably, in addition to the well-established BRS index, nearly all studied mutual information indices exhibited significant differences between the patients' groups and the healthy volunteers ($p < 0.05$). In marked contrast, when it comes to differential diagnostics, only a few studied metrics have shown significant discrepancies between the first and the second patients' groups. In particular, besides the well established BRS index ($p < 0.05$), also Sync ($p < 0.05$) and TDS ($p < 0.005$) mutual information metrics have shown significant differences (see also Figures 4–6 for visual illustration).

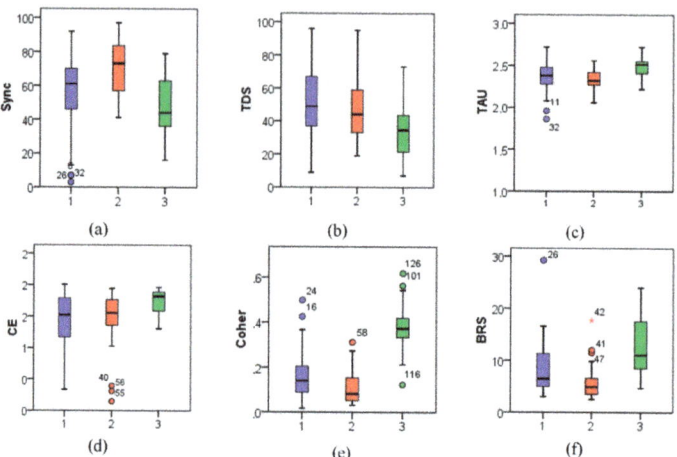

Figure 4. The overall average of the six studied mutual information metrics: (**a**) Sync, (**b**) TDS, (**c**) *TAU*, (**d**) CE, (**e**) *Coher* and (**f**) BRS averaged for the overall duration of the tilt test records. The horizontal axis denotes group number (also denoted by color): 1 (blue)—orthostatic hypotension; 2 (red)—diabetes mellitus with autonomic neuropathy; 3 (green)—healthy volunteers (control group).

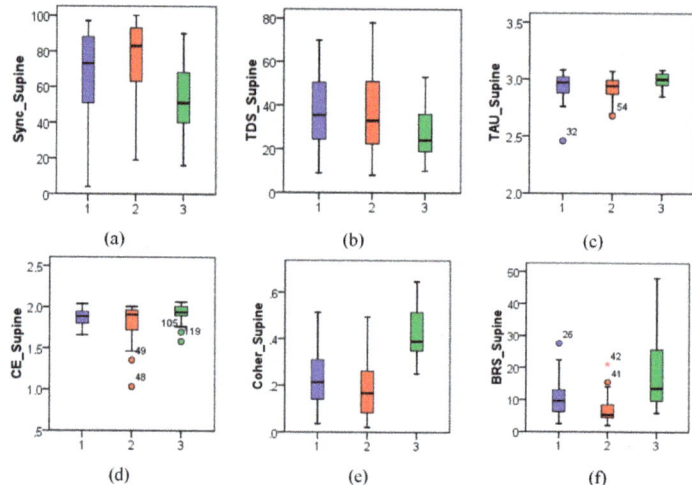

Figure 5. The values of the six studied mutual information metrics: (**a**) Sync, (**b**) TDS, (**c**) *TAU*, (**d**) CE, (**e**) *Coher* and (**f**) BRS obtained during the supine phase of the tilt test. The horizontal axis denotes group number (also denoted by color): 1 (blue)—orthostatic hypotension; 2 (red)—diabetes mellitus with autonomic neuropathy; 3 (green)—healthy volunteers (control group).

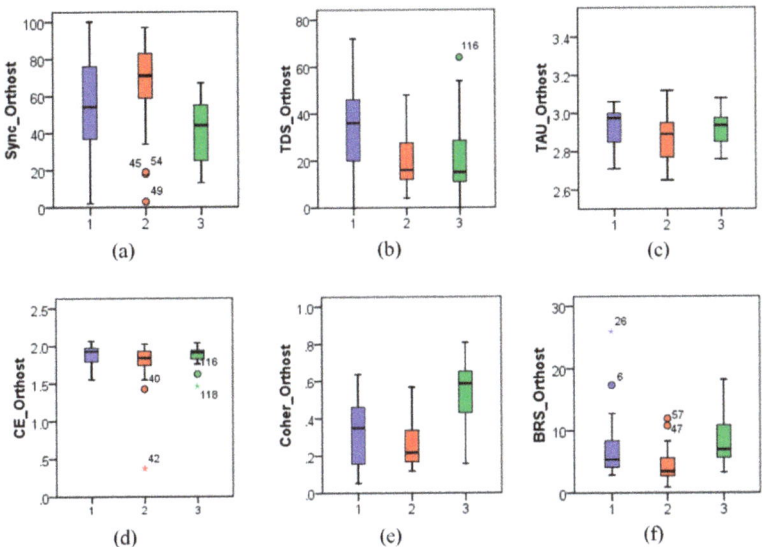

Figure 6. The values of the six studied mutual information metrics: (**a**) Sync, (**b**) TDS, (**c**) *TAU*, (**d**) CE, (**e**) *Coher* and (**f**) BRS obtained during the orthostatic phase of the tilt test. The horizontal axis denotes group number (also denoted by color): 1 (blue)—orthostatic hypotension; 2 (red)—diabetes mellitus with autonomic neuropathy; 3 (green)—healthy volunteers (control group).

For a better representation of the overall response patterns quantified simultaneously by several indicators, Figure 7 also depicts the star-style diagrams summarizing our results after appropriate normalization. Those indices that were not normalized by definition have been rescaled by dividing all values by the corresponding observed maxima.

Significant changes between the supine and the orthostatic phases of the tilt test could be observed for nearly all studied metrics except CE in the diabetes patients' group ($p < 0.05$). However, only BRS and Coher metrics ($p < 0.01$) significantly differed between the supine and the orthostatic test phases both in the control group and in the first patient group with orthostatic hypotension, while no significant differences between the test phases could be observed in other studied indices.

In particular, although BRS reduced significantly in the orthostatic position for the patients with non-diabetic orthostatic hypotension (Group 1), given that their initial *BRS* values (9.59 ± 4.65) (medians ± interquartile ranges are given here and below) in the supine position were only slightly below those in the control group (see Figure 5), they appeared only moderately reduced also in the orthostatic position (5.16 ± 4.12, see also Figure 6). In contrast, in diabetic patients (Group 2) characterized by already low BRS values in the supine position (4.07 ± 2.98, see also Figure 5), its comparable relative reduction in the orthostatic position led to much lower absolute *BRS* values (2.73 ± 1.03, see also Figure 6).

According to these data, one could expect higher blood pressure variability in Group 2, as their autonomic nervous system is less sensitive to the changes in blood pressure, and their feedback response to blood pressure variations is weaker than in Group 1. Surprisingly, patients in Groups 1 and 2 demonstrated rather comparable blood pressure variability (SBP standard deviations were 7.83 ± 4.13 vs. 5.58 ± 3.08 mmHg in the supine and 7.69 ± 4.71 vs. 6.15 ± 1.84 in the orthostatic position, see also Figures 5 and 6, respectively).

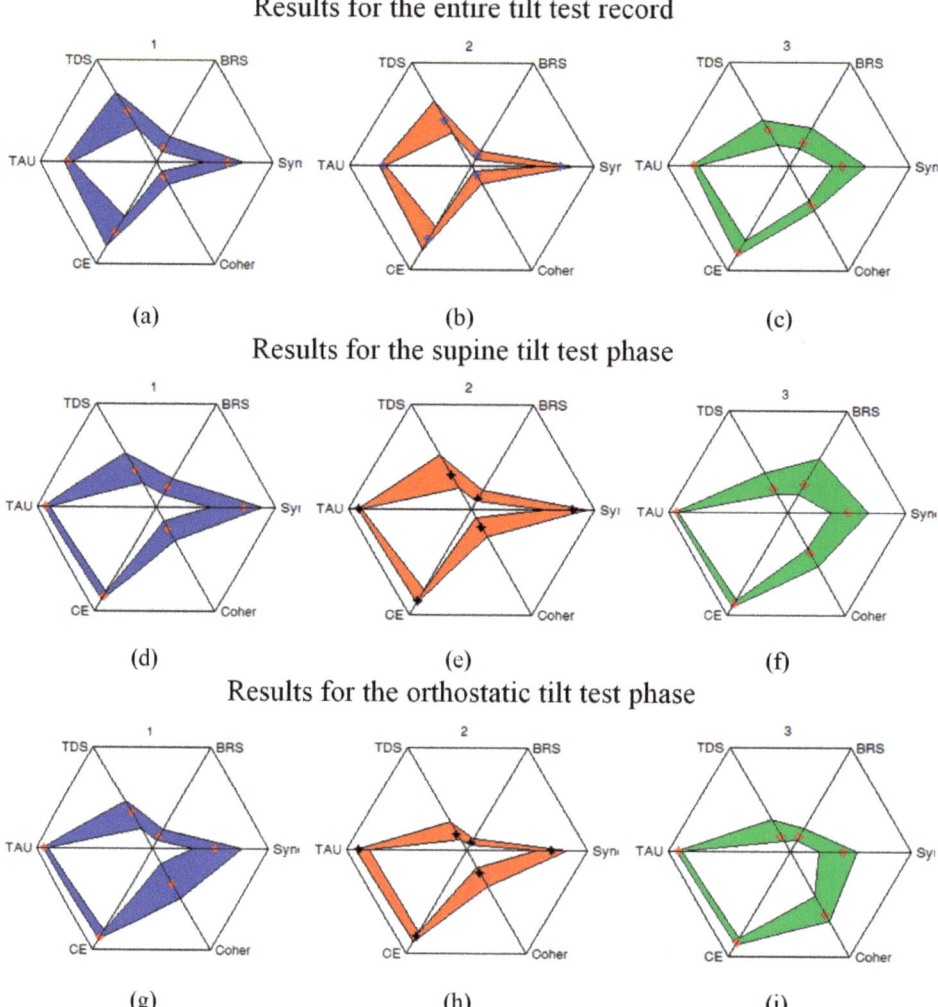

Figure 7. The star-style diagrams representing the overall responses to the tilt test based on a series of studies mutual dynamics indicators averaged for the overall duration of the tilt test records: (**a**) for the orthostatic hypotension patients (group 1), (**b**) for the diabetes patients with autonomic neuropathy (group 2) and (**c**) for the healthy subjects (control group); also for the supine tilt test phase: (**d**) for the orthostatic hypotension patients (group 1), (**e**) for the diabetes patients with autonomic neuropathy (group 2) and (**f**) for the healthy subjects (control group); as well as for the orthostatic tilt test phase: (**g**) for the orthostatic hypotension patients (group 1), (**h**) for the diabetes patients with autonomic neuropathy (group 2) and (**i**) for the healthy subjects (control group). Numbers above the diagrams also indicate the patient/subject groups (also denoted by different colors). The filled areas represent the interquartile ranges (IQR), while small stars within the filled areas represent the medians.

Moreover, the synchronization coefficient Sync demonstrated significant negative correlations with the standard deviation of the systolic blood pressure during the orthostatic phase of the tilt test. This seems to be rather a universal phenomenon, as it is reproduced well not only qualitatively but also quantitatively in all studied groups (Spearman's correlation coefficient $\rho \approx -0.5$, $p < 0.05$). The above indicates that in addition to the established BRS, synchronization patterns between blood pressure and pulse intervals also appear important markers of an adequate cardiovascular response to the orthostatic stress.

4. Discussion

While key indicators of the cardiovascular feedback regulation are characterized by well-known measures such as BRS, some details could not be revealed by using this single indicator, especially when it comes to the differential diagnostics. Despite of the pronounced discrepancies between the BRS values in patients from Groups 1 and 2 (approximately twofold between their median values for both supine and orthostatic tilt-test phases), blood pressure variability did not exhibit any significant discrepancies between the two patient groups, indicating that there might be another contributing factor that also plays an important role.

Such a factor could be revealed and, furthermore, quantified by considering the synchronization coefficient Sync and the time delay stability TDS metrics both indicating the (normalized) total duration of the time fragments when rather synchronous dynamics of blood pressure and pulse intervals could be observed. Our results indicate that in all studied groups under normal conditions both Sync and TDS demonstrate behavior that appears rather *reciprocal* to BRS. This indicates that in patients with reduced BRS the feedback mechanisms are likely being *activated more frequently* this way trying to compensate their lower sensitivity and intensity.

This compensation seems to be rather a universal phenomenon, as the reciprocal character of Sync and TDS vs. BRS can be clearly observed in the comparison between them in different studied groups in both supine and orthostatic positions (see boxplots in Figures 4–6). The compensation hypothesis is further supported by the fact that during *all* tilt test phases in both patients from Group 1 and healthy subjects always negative (although not in all cases statistically significant) correlations between BRS and Sync have been observed. Although, qualitatively, a similar effect could also be observed when considering TDS instead of Sync, it appears less pronounced that could be likely attributed to the lower sensitivity of TDS when compared to Sync, as revealed by a recent computer simulations based study [30].

A prominent exception from the above reciprocal relationship as well as from the corresponding negative correlation pattern could be observed only in diabetes patients likely indicating a breakdown of the above compensatory mechanism at least in some of the patients. Notably, this appears generally in line with recent studies of baroreflex control where significant dependence of the orthostatic baroreflex performance on concomitant conditions such as obesity has been reported [21]. Moreover, in the same study a significantly higher number of baroreflex sequences per given time window (that could also serve as another possible substitute to Sync and/or TDS mutual synchronization metrics) likely indicating more frequent activation of the baroreflex loop could be observed in diabetes patients with concomitant obesity compared to control subjects, although differences between diabetes patients and weight-matched control subjects appeared insignificant. Since we have observed similar effect in our study by using the suggested Sync and TDS indicators, also concomitant conditions such as obesity presumably play a key role in the impairment of baroreflex control not only in terms of its low sensitivity, but also in terms of its timely activation in response to blood pressure variations, although more detailed investigations are required to further elucidate the key factors that influence this possible compensation breakdown.

Therefore, we believe that proposed additional indicators could be useful for the improvement of the differential diagnostics of blood pressure regulation efficacy and also lead to a deeper insight into the involved concomitant factors this way also aiming at the improvement of the mathematical models representing the underlying feedback control mechanisms.

Finally, the proposed complementary mutual behavior indicators might appear potentially useful for the analysis of other physiological signals as well as for the quantification of the alternative stress test responses other that the tilt test, for example, in stress detection studies in daily life scenarios (for recent examples of relevant investigations see e.g., [40,41]). However, for further practical utilization of these complementary indicators in other differential diagnostic scenarios, design of dedicated prediction tools based on, e.g., star-style diagram pattern recognition and/or shape analysis with anomaly detection [42] or multivariate regression models with decision-making procedure based on the analysis of appropriately weighted combination of several complementary indices (see, e.g., [43] for a recent example) is required.

5. Conclusions

To summarize, six different metrics of mutual coupling of simultaneously registered signals representing blood pressure and pulse interval dynamics have been considered in this study. Stress test response patterns represented by the reaction of the recorded signals to the external input by tilting the body into the upright position have been analyzed. While nearly *all* studied metrics significantly differed between patients and healthy subjects, only a few of them appeared informative for the differential diagnostics of patients with autonomic disorders of different etiology and severity. Besides the widely used BRS index, also Sync and TDS mutual information metrics representing the temporal fractions of the analyzed signal records exhibiting rather synchronous dynamics exhibited significant differences. While BRS characterizes the *intensity* of the pulse intervals response to the blood pressure changes during observed feedback responses, both Sync and TDS likely indicate how often such responses are *activated* in the first place. Our results indicate, that in most cases, BRS is typically *reciprocal* to both Sync and TDS suggesting that *low intensity* of the feedback responses characterized by low BRS is rather compensated by their *more frequent activation* indicated by higher Sync and TDS. A notable exception can be observed in diabetes patients with autonomic neuropathy, where a likely breakdown of this compensation could be observed.

Author Contributions: Conceptualization, M.I.B. and O.V.M.; formal analysis, N.S.P.; investigation, S.A.P. and M.I.B.; project administration, O.A.M. and M.I.B.; resources, O.V.M.; software, N.S.P.; supervision, M.I.B.; writing—original draft, M.I.B.; writing—review and editing, O.A.M., O.V.M. and M.I.B. All authors have read and agreed to the published version of the manuscript.

Funding: We would like to acknowledge the financial support of this work by the Ministry of Science and Education of the Russian Federation in the framework of the basic state assignment No. 2.5475.2017/6.7.

Conflicts of Interest: The authors declare no conflict of interest. The funders had no role in the design of the study; in the collection, analyses, or interpretation of data; in the writing of the manuscript, or in the decision to publish the results.

References

1. Bertinieri, G.; Di Rienzo, M.; Cavallazzi, A.; Ferrari, A.U.; Pedotti, A.; Mancia, G. A new approach to analysis of arterial baroreflex. *J. Hypertens.* **1985**, *3* (Suppl. 3), 79–81.
2. Malberg, H.; Wessel, N.; Hasart, A.; Osterziel, K.J.; Voss, A. Advanced analysis of spontaneous baroreflex sensitivity, blood pressure and heart rate variability in patients with dilated cardiomyopathy. *Clin. Sci.* **2002**, *102*, 465–473. [CrossRef] [PubMed]
3. De Boer, R.W.; Karemaker, J.M.; Strackee, J. Relations between short-term blood pressure fluctuations and heart rate variability in resting subjects: A spectral analysis approach. *Med. Biol. Eng. Comp.* **1985**, *23*, 352–364. [CrossRef] [PubMed]
4. Robbe, H.W.; Mulder, L.J.; Ruddel, H.; Langewitz, W.A.; Veldman, J.B.; Mulder, G. Assessment of Baroreceptor Reflex Sensitivity by Means of Spectral Analysis. *Hypertension* **1987**, *10*, 538–543. [CrossRef] [PubMed]
5. Cerutti, S.; Baselli, G.; Civardi, S.; Furlan, R.; Lombardi, F.; Malliani, A.; Merri, M.; Pagani, M. Spectral analysis of heart rate and blood pressure variability signals for physiological and clinical purposes. *Comput. Cardiol.* **1987**, *14*, 435–438.
6. Bogachev, M.I.; Mamontov, O.V.; Konradi, A.O.; Uljanitski, Y.D.; Kantelhardt, J.W.; Schlyakhto, E.V. Analysis of blood pressure–heart rate feedback regulation under non-stationary conditions: Beyond baroreflex sensitivity. *Physiol. Meas.* **2009**, *630*, 631–645. [CrossRef] [PubMed]

7. Parati, G.; Omboni, S.; Fratolla, A.; Di Rienzo, M.; Zanchetti, A.; Mancia, G. Dynamic evaluation of baroreflex in ambulant subjects. In *Blood Pressure and Heart Rate Variablity*; Mancia, G., Parati, G., Pedotti, A., Zanchetti, A., Di Rienzo, M., Eds.; IOS Press: Amsterdam, The Netherlands, 1992; pp. 123–137.
8. La Rovere, M.T.; Bigger, J.T., Jr.; Marcus, F.I.; Mortara, A.; Schwartz, P.J. Baroreflex sensitivity and heart-rate variability in prediction of total cardiac mortality after myocardial infarction. *Lancet* **1998**, *351*, 478–484. [CrossRef]
9. De Ferrari, G.M.; Sanzo, A.; Bertoletti, A.; Specchia, G.; Vanoli, E.; Schwartz, P.J. Baroreflex sensitivity predicts long-term cardiovascular mortality after myocardial infarction even in patients with preserved left ventricular function. *J. Am. Coll. Cardiol.* **2007**, *50*, 2285–2290. [CrossRef]
10. La Rovere, M.T.; Pinna, G.D.; Maestri, R.; Robbi, E.; Caporotondi, A.; Guazzotti, G.; Sleight, P.; Febo, O. Prognostic implications of baroreflex sensitivity in heart failure patients in the beta-blocking era. *J. Am. Coll. Cardiol.* **2009**, *53*, 193–199. [CrossRef]
11. La Rovere, M.T.; Pinna, G.D.; Hohnloser, S.H.; Marcus, F.I.; Mortara, A.; Nohara, R.; Bigger, J.T., Jr.; Camm, A.J.; Schwartz, P.J. Baroreflex sensitivity and heart rate variability in the identification of patients at risk for life-threatening arrhythmias: Implications for clinical trials. *Circulation* **2001**, *103*, 2072–2077. [CrossRef]
12. La Rovere, M.T.; Pinna, G.D.; Maestri, R.; Sleigh, P. Clinical value of baroreflex sensitivity. *Neth. Heart J.* **2013**, *21*, 61–63. [CrossRef] [PubMed]
13. Mathias, C.J.; Bannister, R. (Eds.) *Autonomic Failure: A Textbook of Clinical Disorders of the Autonomous Nervous System*; Oxford University Press: Oxford, UK, 2013.
14. Steptoe, A.; Vögele, C. Cardiac baroreflex function during postural change assessed using non-invasive spontaneous sequence analysis in young men. *Cardiovasc. Res.* **1990**, *24*, 627–632. [CrossRef] [PubMed]
15. Kardos, A.; Rudas, L.; Simon, J.; Gingl, Z.; Csanady, M. Effect of postural changes on arterial baroreflex sensitivity assessed by the spontaneous sequence method and Valsalva manoeuvre in healthy subjects. *Clin. Auton. Res.* **1997**, *7*, 143–148. [CrossRef]
16. James, M.A.; Potter, J.F. Orthostatic blood pressure changes and arterial baroreflex sensitivity in elderly subjects. *Age Ageing* **1999**, *28*, 522–530. [CrossRef]
17. Mattace-Raso, F.U.; van den Meiracker, A.H.; Bos, W.J.; van der Cammen, T.J.; Westerhof, B.E.; Elias-Smale, S.; Reneman, R.S.; Hoeks, A.P.; Hofman, A.; Witteman, J.C. Arterial stiffness, cardiovagal baroreflex sensitivity and postural blood pressure changes in older adults: The Rotterdam Study. *J. Hypertens.* **2007**, *25*, 1421–1426. [CrossRef]
18. Fadel, P.J.; Raven, P.B. Human investigations into the arterial and cardiopulmonary baroreflexes during exercise. *Exp. Physiol.* **2011**, *97*, 39–50. [CrossRef]
19. Schwartz, C.E.; Stewart, J.M. The arterial baroreflex resets with orthostasis. *Front. Physiol.* **2012**, *3*, 461. [CrossRef]
20. Dampney, R.A.L. Resetting the baroreflex control of sympathetic vasomotor activity during natural behaviors: Description and conceptual model of central mechanisms. *Front. Neurosci.* **2017**, *11*, 461. [CrossRef]
21. Holwerda, S.W.; Vianna, L.C.; Restaino, R.M.; Chaudhary, K.; Young, C.N.; Fadel, P.J. Arterial baroreflex control of sympathetic nerve activity and heart rate in patients with type 2 diabetes. *Am. J. Physiol. Heart Circ. Physiol.* **2016**, *311*, H1170–H1179. [CrossRef]
22. Bauer, A.; Kantelhardt, J.W.; Barthel, P.; Schneider, R.; Mkikallio, T.; Ulm, K.; Hnatkova, K.; Schmig, A.; Huikuri, H.; Bunde, A.; et al. Deceleration capacity of heart rate as a predictor of mortality after myocardial infarction: Cohort study. *Lancet* **2006**, *367*, 1674–1681. [CrossRef]
23. Bisognano, J.D.; Bakris, G.; Nadim, M.K.; Sanchez, L.; Kroon, A.A.; Schafer, J.; De Leeuw, P.W.; Sica, D.A. Baroreflex activation therapy lowers blood pressure in patients with resistant hypertension: Results from the double-blind, randomized, placebo-controlled rheos pivotal trial. *J. Am. Coll. Cardiol.* **2011**, *58*, 765–773. [CrossRef] [PubMed]
24. Bartsch, H.; Bartsch, C.; Mecke, D.; Lippert, T.H. Seasonality of pineal melatonin production in the rat: Possible synchronization by the geomagnetic field. *Chronobiol. Int.* **1994**, *11*, 21–26. [CrossRef] [PubMed]
25. Vosko, A.M.; Colwell, C.S.; Avidan, A.Y. Jet lag syndrome: Circadian organization, pathophysiology, and management strategies. *Nat. Sci. Sleep* **2010**, *2*, 187–198. [PubMed]
26. Ramkisoensing, A.; Meijer, J.H. Synchronization of Biological Clock Neurons by Light and Peripheral Feedback Systems Promotes Circadian Rhythms and Health. *Front. Neurol.* **2015**, *6*, 128. [CrossRef] [PubMed]

27. Bartsch, R.P.; Kantelhardt, J.W.; Penzel, T.; Havlin, S. Experimental evidence for phase synchronization transitions in the human cardiorespiratory system. *Phys. Rev. Lett.* **2007**, *98*, 054102. [CrossRef]
28. Bartsch, R.P.; Schumann, A.Y.; Kantelhardt, J.W.; Penzel, T.; Ivanov, P.C.H. Phase transitions in physiologic coupling. *Proc. Nat. Acad. Sci. USA* **2012**, *109*, 10181–10186. [CrossRef]
29. Bartsch, R.P.; Liu, K.K.L.; Bashan, A.; Ivanov, P.C.H. Network physiology: How organ systems dynamically interact. *PLoS ONE* **2015**, *10*, e0142143. [CrossRef]
30. Pyko, N.S.; Pyko, S.A.; Markelov, O.A.; Karimov, A.I.; Butusov, D.N.; Zolotukhin, Y.V.; Uljanitski, Y.D.; Bogachev, M.I. Assessment of cooperativity in complex systems with non-periodical dynamics: Comparison of five mutual information metrics. *Phys. A Stat. Mech. Its Appl.* **2018**, *503*, 1054–1072. [CrossRef]
31. Vinik, A.I.; Maser, R.E.; Mitchell, B.D.; Freeman, R. Diabetic autonomic neuropathy. *Diabetes Care* **2005**, *26*, 1553–1579. [CrossRef]
32. Spallone, V.; Ziegler, D.; Freeman, R.; Bernardi, L.; Frontoni, S.; Pop-Busui, R.; Stevens, M.; Kempler, P.; Hilsted, J.; Tesfaye, S.; et al. Cardiovascular autonomic neuropathy in diabetes: Clinical impact, assessment, diagnosis, and management. *Diabetes Metab. Res. Rev.* **2011**, *27*, 639–653. [CrossRef]
33. Brignole, M.; Alboni, P.; Benditt, D.G.; Bergfeldt, L.; Blanc, J.-J.; Thomsen, P.E.B.; van Dijk, J.G.; Fitzpatrick, A.; Hohnloser, S.; Janousek, J.; et al. 2004 Guidelines on Management (diagnosis and treatment) of syncope-update 2004. The task force on syncope, European society of Cardiology. *Europace* **2004**, *6*, 467–537. [PubMed]
34. Novak, P. Quantitative Autonomic Testing. *J. Vis. Exp.* **2011**, *53*, 2502. [CrossRef] [PubMed]
35. Markelov, O.A.; Bogachev, M.I.; Mamontov, O.V.; Katinas, G.S. An integrated algorithmic and software solution for biological rhythms analysis: Application to long-term data series. In Proceedings of the 2014 Mechanical Engineering, Automation and Control. Systems (MEACS), Tomsk, Russia, 16–18 October 2014.
36. Pyko, N.S.; Pyko, S.A.; Markelov, O.A.; Bogachev, M.I. Systolic blood pressure and pulse intervals synchronization. In Proceedings of the 2015 IEEE NW Russia Young Researchers in Electrical and Electronic Engineering Conference, St. Petersburg, Russia, 2–4 February 2015; pp. 341–344.
37. Bogachev, M.I.; Markelov, O.A.; Pyko, N.S.; Pyko, S.A. Blood pressure—Heart rate synchronization coefficient as a complementary indicator of baroreflex mechanism efficiency. In Proceedings of the 2015 Soft Computing and Measurements (SCM), St. Petersburg, Russia, 19–21 May 2015; pp. 173–175.
38. Julien, C. The enigma of Mayer waves: Facts and models. *Cardiovasc. Res.* **2006**, *70*, 12–21. [CrossRef] [PubMed]
39. Landau, S.; Everitt, B.S. *A Handbook of Statistical Analyses Using SPSS*, 1st ed.; Chapman and Hall/CRC: Boca Raton, FL, USA, 2003.
40. Hussain, G.; Jabbar, M.S.; Bae, S.; Cho, J.D. Stress detection of the students studying in university using smartphone sensors SPWID. In Proceedings of the Fourth International Conference on Smart Portable, Wearable, Implantable and Disability-oriented Devices and Systems, Barcelona, Spain, 22–26 July 2018; pp. 30–34.
41. Can, Y.S.; Arnrich, B.; Ersoy, C. Stress detection in daily life scenarios using smart phones and wearable sensors: A survey. *J. Biomed. Inf.* **2019**, *92*, 103139. [CrossRef] [PubMed]
42. Krasichkov, A.S.; Grigoriev, E.B.; Bogachev, M.I.; Nifontov, E.M. Shape anomaly detection under strong measurement noise: An analytical approach to adaptive thresholding. *Phys. Rev. E* **2015**, *92*, 042927. [CrossRef]
43. Bogachev, M.I.; Kayumov, A.R.; Markelov, O.A.; Bunde, A. Statistical prediction of protein structural, localization and functional properties by the analysis of its fragment mass distributions after proteolytic cleavage. *Sci. Rep.* **2016**, *6*, 22286. [CrossRef]

© 2019 by the authors. Licensee MDPI, Basel, Switzerland. This article is an open access article distributed under the terms and conditions of the Creative Commons Attribution (CC BY) license (http://creativecommons.org/licenses/by/4.0/).

Article
Wood Defect Detection Based on Depth Extreme Learning Machine

Yutu Yang [1], Xiaolin Zhou [2], Ying Liu [1,*], Zhongkang Hu [3] and Fenglong Ding [1]

1. College of Mechanical & Electronic Engineering, Nanjing Forestry University, Nanjing 210037, China; yangyutu@njfu.edu.cn (Y.Y.); dfl@njfu.edu.cn (F.D.)
2. School of Artificial Intelligence, Hezhou University, Hezhou 542899, China; zhouxiaolinzxl@hotmail.com
3. Nanjing Fujitsu Nanda Software Technology Co., Ltd., Nanjing 210012, China; huzk.fnst@cn.fujitsu.com
* Correspondence: liuying@njfu.edu.cn

Received: 18 September 2020; Accepted: 22 October 2020; Published: 24 October 2020

Abstract: The deep learning feature extraction method and extreme learning machine (ELM) classification method are combined to establish a depth extreme learning machine model for wood image defect detection. The convolution neural network (CNN) algorithm alone tends to provide inaccurate defect locations, incomplete defect contour and boundary information, and inaccurate recognition of defect types. The nonsubsampled shearlet transform (NSST) is used here to preprocess the wood images, which reduces the complexity and computation of the image processing. CNN is then applied to manage the deep algorithm design of the wood images. The simple linear iterative clustering algorithm is used to improve the initial model; the obtained image features are used as ELM classification inputs. ELM has faster training speed and stronger generalization ability than other similar neural networks, but the random selection of input weights and thresholds degrades the classification accuracy. A genetic algorithm is used here to optimize the initial parameters of the ELM to stabilize the network classification performance. The depth extreme learning machine can extract high-level abstract information from the data, does not require iterative adjustment of the network weights, has high calculation efficiency, and allows CNN to effectively extract the wood defect contour. The distributed input data feature is automatically expressed in layer form by deep learning pre-training. The wood defect recognition accuracy reached 96.72% in a test time of only 187 ms.

Keywords: wood defect; CNN; ELM; genetic algorithm; detection

1. Introduction

Today's wood products are manufactured under increasingly stringent requirements for surface processing. In developed countries such as Sweden and Finland with developed forest resources, the comprehensive use rate of wood is as high as 90%. In sharp contrast, the comprehensive use rate of wood in China is less than 60%, causing a serious waste of resources. With China's rapid economic development, people are increasingly pursuing a high-quality life, which will inevitably lead to an increase in demand for wood and wood products, such as solid wood panels, wood-based panels, paper and cardboard, and other consumption levels are among the highest in the world. The existing wood storage capacity and processing level make it difficult to meet the rapid growth demand. The lack of wood supply and the low use rate have led to the limited development of China's wood industry. Therefore, it is necessary to comprehensively inspect the processing quality of logs and boards to improve the use rate of wood and the quality of wood products.

The nondestructive testing of wood can accurately and quickly make judgments on the physical properties and growth defects in wood, and nondestructive wood testing and automation can be realized. In recent years, the combined application of computer technology along with detection and control theory has made great progress in the detection of wood defects. In the nondestructive testing of wood surfaces, commonly used traditional methods include laser testing [1,2], ultrasonic testing [3–5], acoustic emission technology [6,7], etc. Computer-aided techniques are a common approach to surface processing, as they are efficient and have a generally high recognition rate [8,9]. Deep learning was first proposed by Hinton in 2006; in 2012, scholars adopted an AlexNet network based on deep learning to achieve computer vision recognition accuracy of up to 84.7%. Deep learning prevents dimensionality in layer initialization and represents a revolutionary development in the field of machine learning [10,11]. More and more scholars are applying deep learning networks in wood nondestructive testing. He [12] et al. used a linear array CCD camera to obtain wood surface images, and proposed a hybrid total convolution neural network (Mix-FCN) for the recognition and location of wood defects; however, the network depth was too deep and required too much calculation. Hu [13] and Shi [14] used the Mask R-CNN algorithm in wood defect recognition, but they used a combination of multiple feature extraction methods, which resulted in a very complex model. However, the current deep learning algorithms still have problems such as inaccurate defect location, incomplete defect contour and boundary information in the wood defect detection process. To solve the above problems and effectively meet the needs of wood processing enterprises for wood testing, we carry out the research of this article.

The innovations of this article are: (1) Simple pre-processing of wood images using nonsubsampled shear wave transform (NSST), reducing the complexity and computational complexity of image processing, as the input of convolutional neural network; (2) application of a simple linear iterative clustering (SLIC) algorithm to enhance and improve the convolutional neural network to obtain a super pixel image with a more complete boundary contour; (3) use of genetic algorithm to improve extreme learning machine and classified the obtained image features. Through the above method, the accuracy of defect detection is improved, and the recognition time is truncated to establish an innovative machine-vision-based wood testing technique.

2. Materials and Methods

2.1. Wood Defect Original Image Dataset

According to the different processes and causes of solid wood board defects, they are divided into biohazard defects, growth defects and processing defects. Among them, growth defects and biohazard defects are natural defects, which have certain shape and structure characteristics, and are also an important basis for wood grade classification. Generally speaking, solid wood board growth defects and biohazard defects can be divided into: dead knots, live knots, worm holes, decay, etc. The original data set used in the experiment in this article is derived from the wood sampling image in the 948 project of the State Forestry Administration (the introduction of the laser profile and color integrated scanning technology for solid wood panels). When scanning to obtain wood images, the scanning speed of the scanner is 170 Hz–5000 Hz; Z direction resolution is 0.055 mm–0.200 mm; X direction resolution is 0.2755 mm–0.550 mm; and color pixel resolution can reach 1 mm × 0.5 mm. The data set includes 5000 defect maps of pine, fir, and ash. The bit depth of each image is 24, and the specified size is at the 100*100 pixel level. Part of the defect image is shown in Figure 1.

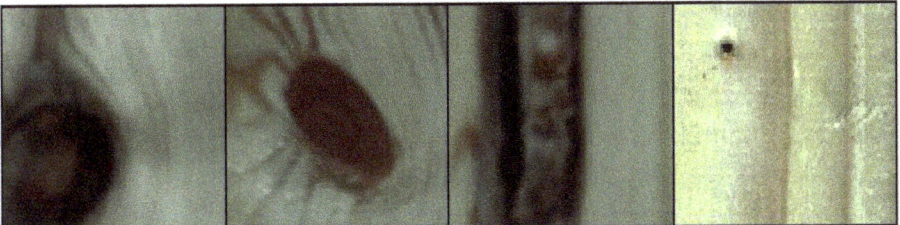

Figure 1. Common defects of solid wood such as dead-knot, live-knot and decay.

2.2. Optimized Convolution Neural Network

This paper proposes an optimized algorithm which uses NSST to preprocess the images followed by the CNN to extract defect features from wood images as a preliminary CNN model. The simple linear iterative clustering (SLIC) super-pixel segmentation algorithm is used to analyze the wood images by super-pixel clustering, which allows the defects in wood images and local information regarding defects and cracks to be efficiently located. The obtained information is fed back to the initial model, which enhances the original CNN.

2.2.1. Structure and Characteristics of Convolution Neural Networks

The CNN is an artificial neural network algorithm with multi-layer trainable architecture [15]. It generally consists of an input layer, excitation layer, pool layer, convolution layer, and full connection layer. CNNs have many advantages in terms of image processing applications. (1) Feature extraction and classification can be combined into the same network structure and synchronized training can be achieved, and the algorithm is fully adaptive. (2) When the image size is larger, the deep feature information can be extracted better. (3) Its unique network structure has strong adaptability to the local deformation, image rotation, image translation, and other changes in the input image. In this study, each pixel in the wood image was convoluted and the defect feature was extracted by exploiting these CNN characteristics. The CNN network skeleton used in this article is shown in Figure 2.

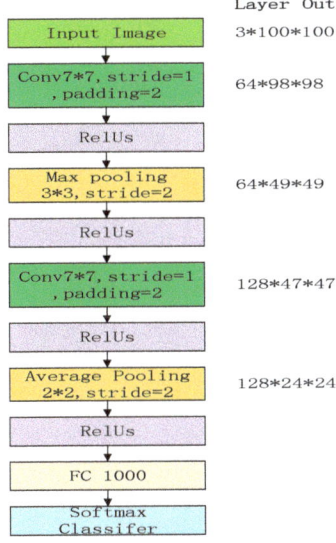

Figure 2. CNN network skeleton.

2.2.2. Non-Subsampled Shearlet Transform (NSST)

The NSST can represent signals sparsely and optimally, but it also has a strong direction-sensitivity [16–18]. Therefore, using NSST to preprocess wood images can preserve the defects feature of wood images. Redundancies in the wood image information are reduced in addition to the complexity and computation of image processing with the depth learning method.

2.2.3. Simple Linear Iterative Clustering (SLIC)

The CNN uses a matrix form to represent an image to be processed, so the spatial organization relationship between pixels is not considered—this affects the image segmentation and obscures the boundary of the defective region of the wood image. The SLIC algorithm can generate relatively compact super-pixel image blocks after processing a gray or color image. The generated super-pixel image is compact between pixels, and the edge contour of the image is clear. To this effect, the SLIC extracts a relatively accurate contour to supplement the feature contour. The SLIC also works with relatively few initial parameters. Only the number of hyper pixels needed to segment the image must be set. The algorithm is simple in principle, and has a small calculation range and rapid running speed. By 2015, the parallel execution speed had reached 250 FPS; it is now the fastest super-pixel segmentation method available [19].

2.2.4. Feature Extraction

The optimized CNN model proposed in this paper was designed for wood surface feature extraction. Knots were used as example wood defects (Figure 3a) to test feature extraction via the following operations. The input image Figure 3a was directly processed by CNN algorithm to obtain image Figure 3b, which presented local irregularity and nonsmooth edges in the contour after enlargement [20]. The SLIC algorithm was used to process the input image (Figure 3a) followed by longitudinal convolution (Figure 3d). The image shown in Figure 3h was obtained after edge removal and fusion processing. The defect contour features of Figure 3h are substantially clearer compared to Figure 3b because the segmentation of wood images using CNN, which is expressed in pixels as a matrix without considering the spatial organization relationship between pixels, affects the end image segmentation results. The SLIC algorithm instead extracts the wood defect boundary and contour information from the original image and feeds back the information to the initial segmentation results of the CNN model.

The above process reduces the redundancy of local image information in addition to the complexity and computation of the image processing. The pixel-level CNN model method does not accurately reveal the boundary of the defective region of wood image, but instead indicates only its general position. SLIC can extract a relatively accurate contour to supplement it and optimize the initial CNN model. To this effect, the proposed SLIC algorithm-based method improves the defect feature extraction of wood images over CNN alone.

The input image (Figure 3a) was processed by the NSST algorithm to obtain the image shown in Figure 3e, then Figure 3f was obtained using the SLIC algorithm. Vertical convolution was carried out to obtain the image shown in Figure 3g. Figure 3i was obtained after edge removal and fusion processing. Consider Figure 3i compared to Figure 3h: although the wood defect contour feature extraction effects are not obvious, using NSST to preprocess the image reduces environmental interference and training depth to markedly decrease the computation and complexity of the image processing.

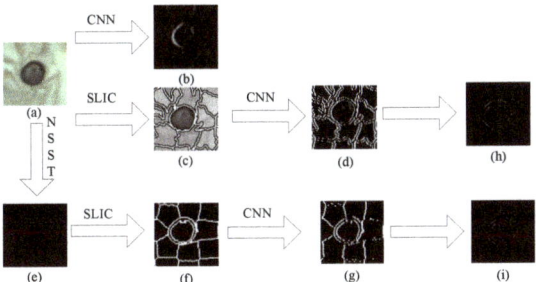

Figure 3. Contrast diagram of optimized CNN feature extraction effects.

Based on the above analysis, this paper decided to use the wood image processing frame shown in Figure 4 to obtain the wood defect feature map.

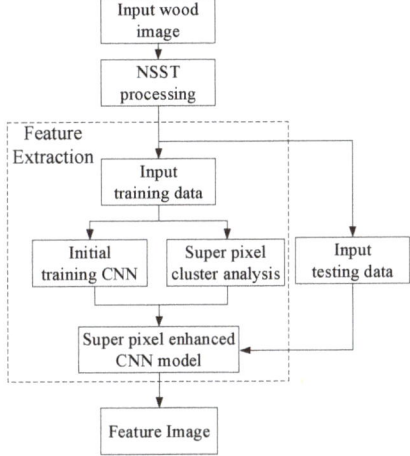

Figure 4. Wood image processing frame.

2.3. Extreme Learning Machine (ELM)

Depth learning is commonly used in target recognition and defect detection applications due to its excellent feature extraction capability. The CNN is highly time-consuming due to the necessity of iterative pre-training and fine-tuning stages; the hardware requirements for more complex engineering applications are also high. The deep CNN structure has a large quantity of adjustable free parameters, which makes its construction highly flexible. On the other hand, it lacks theoretical guidance and is overly reliant on experience, so its generalization performance is dubious. In this study, we integrated the ELM into a depth extreme learning machine (Figure 5) to improve the training efficiency of the deep convolution network. The proposed method extracts wood defects by using an optimized CNN and ELM classifier to exploit the excellent feature extraction ability of the deep network and fast training of ELM simultaneously.

The ELM algorithm differs from traditional pre-feedback neural network training learning. Its hidden layer does not need to be iterated, and input weights and hidden layer node biases are set randomly to minimize training error. The output weights of the hidden layer are determined by the algorithm [21–23]. The ultimate learning machine is based on the proved ordinary extreme theorem and interpolation theorem, under which when the hidden layer activation function of a single hidden layer feedforward neural network is infinitely differentiable, its learning ability is independent of the hidden layer parameters and is only related to the current network structure. When the input weights and hidden layer node offsets are randomly assigned to obtain the appropriate network structure, the ELM has universal approximation capability. The network input weights and hidden layer node offsets can be randomly assigned by approximating any continuous function. Under the premise of network hidden layer activation function infinite differentiability, the output weights of the network can be calculated via the least square method. The network model that can approximate the function can be established, and the corresponding neural network functions such as classification, regression, and fitting can be realized.

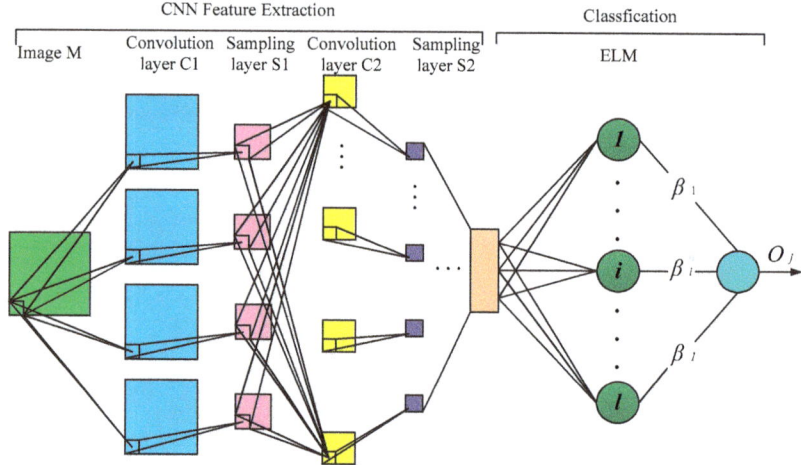

Figure 5. Contrast diagram of optimized CNN feature extraction effects.

This paper mainly centers on the classification function of ELM, which serves to select a relatively simple single hidden layer neural network as the classifier. The traditional neural network algorithm needs many iterations and parameters, learns slowly, has poor expansibility, and requires intensive manual interventions. The ELM used here requires no iterations, the learning speed is relatively fast, the input weights and biases are generated randomly, the follow-up does not need to be set, and relatively few manual interventions are required. In the large sample database, the recognition rate of ELM is better than that of the support vector machine (SVM). For these reasons, we use ELMs as classifiers to enhance recognition efficiency and performance [24,25].

The ELM algorithm introduced above is the main classification method for wood defect feature recognition in this paper. However, in the ELM network structure, the input weights and the threshold of hidden layer nodes are given randomly. For the ELM structure with the same number of hidden layer neurons, the performance of the network is very different, which makes the classification performance of the network more unstable. The genetic algorithm (GA) simulates Darwinian evolutionary theory to optimize the initial weights and threshold of ELM by eliminating less-fit weights and thresholds.

Figure 6a,c show the variation curves of GA-ELM population fitness functions under Radbas, Hardlim, and Sigmoid excitation functions, respectively. Smaller fitness values indicate higher accuracy; the Sigmoid incentive function has the best network effect.

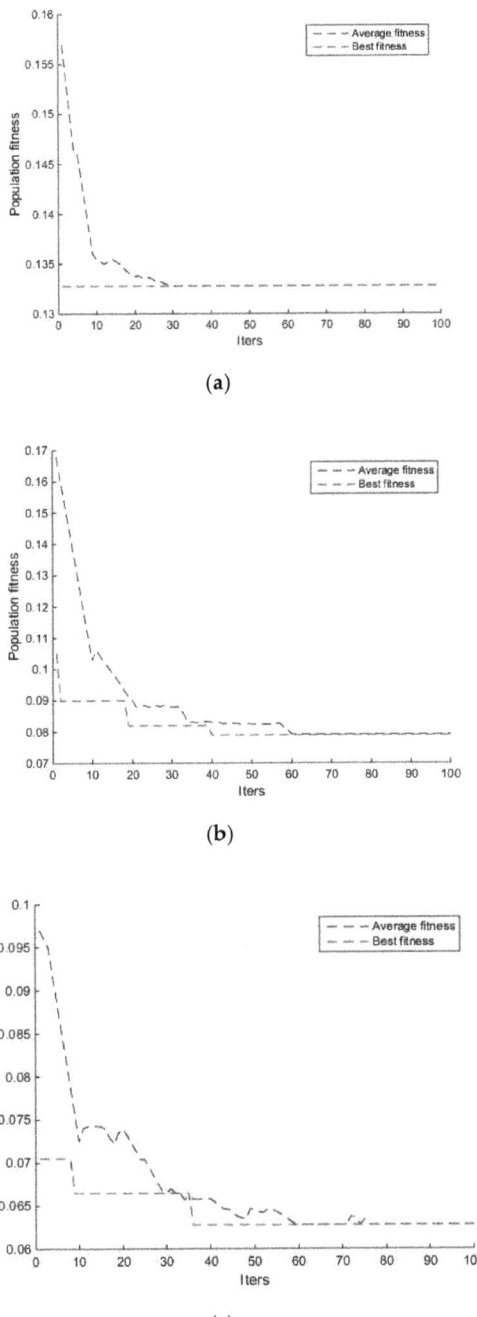

Figure 6. Variation curves of driving function population fitness; (**a**) Radbas driving function; (**b**) Hardlim driving function; (**c**) Sigmoid driving function.

The classification accuracy of GA-ELM and ELM under different excitation functions is also shown in Table 1. We found that the classification accuracy of Sigmoid and Radbas excitations were similar, and the Hardlim excitation function was an exception. The accuracy of ELM and GA-ELM was highest when the Sigmoid function was used as the activation function. The accuracy of GA-ELM reached 95.93%, which is markedly better than that of an unoptimized ELM. The GA optimized ELM network required fewer hidden layer nodes and showed higher test accuracy as well.

Table 1. Classification accuracy of GA-ELM and ELM under different excitation functions.

Excitation Function	Classification Accuracy Rate (%)		Number of Hidden Node	
	GA-ELM	ELM	GA-ELM	ELM
Sigmoid	95.93	93.85	70	90
Hardlim	93.98	91.37	90	170
Radbas	95.37	93.36	110	200

In summary, an improved depth extreme learning machine was constructed in this study by combining the optimized GA-ELM classifier with the optimized CNN feature extraction. It is referred to from here on as "D-ELM".

3. Experimentation

3.1. Experimental Parameters

Table 2 shows the computer-related parameters and software platform used by the experimental system, including CPU model, main frequency and memory size.

Table 2. Experimental parameters.

Items	Type
Server	Intel(R)Xeon(R)CPU E5-4603v2
Quantity of physical CPUs	2
Main frequency of CPUs	2.20 GHz
Memory	16 GB
Experimental platform	Anaconda 3.5, Microsoft Visual Studio 2017

3.2. Empirical Method and Result

The specific experimental process is shown in Figure 7. First, we preprocessed 5000 original images via NSST and randomly selected 4000 images for training. Second, for each pixel in each image, the neighborhood subgraph was taken as the input of CNN, and a total of 40,000,000 samples were obtained as the experimental training set to train the network model. The remaining 10,000,000 samples were used as test images to evaluate the algorithm. The features extracted from the test samples were input into the ELM network classifier, the number of hidden layer nodes of the extreme learning machine was set to 100, then the accuracy and stability of the feature extraction method was statistically analyzed. We found that when the number of iterations exceeds 3500, the loss function is around 0.2 and the convergence performance is acceptable.

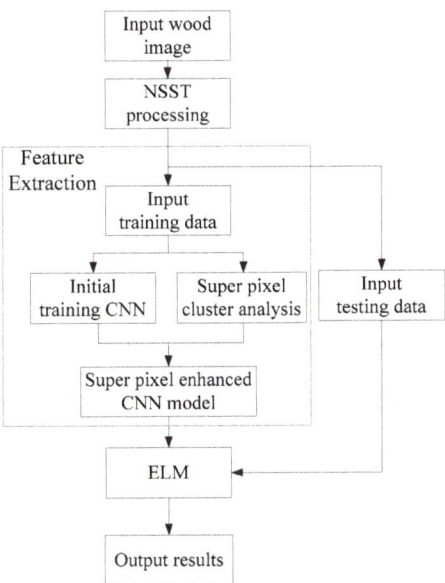

Figure 7. Flowchart of wood defect feature extraction and classification process.

Figure 8a shows the relationship between the training loss value and the number of iterations. Although the training loss value fluctuates a little during the iteration process, it shows a downward trend as a whole. When the iteration is completed, the training loss value was around 0.2. Figure 8b shows a graph of accuracy. When the number of iterations was 1500, the accuracy of the proposed algorithm reached 90%. Accuracy continued to increase as iteration quantity increased until reaching a maximum of about 98%.

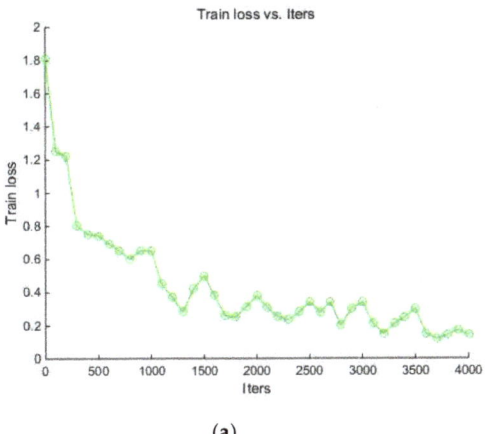

(a)

Figure 8. *Cont.*

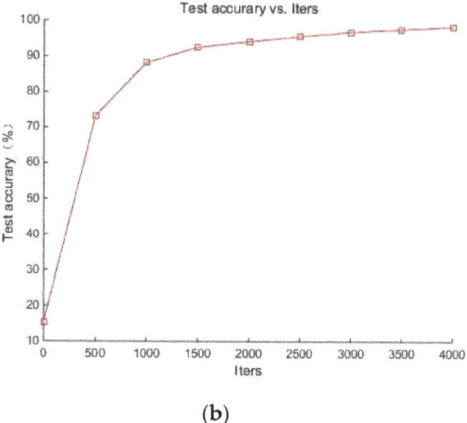

(b)

Figure 8. Relationship between loss function, accuracy, iteration number: (**a**) Relationship between loss function and iteration; (**b**) accuracy graph.

Figure 9 shows our final recognition effect on the test set. We surround the identified wood defects with different colored rectangular borders

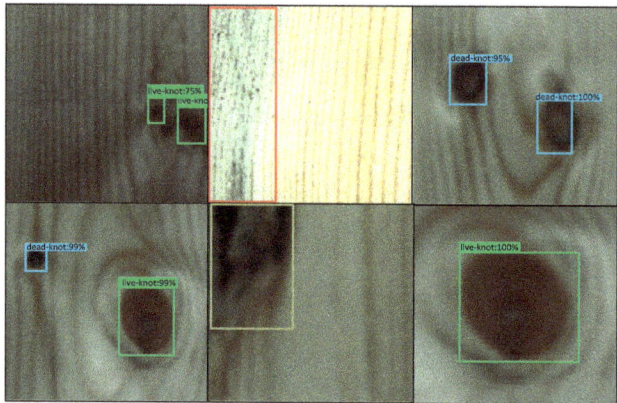

Figure 9. Recognition results based on based on deep learning.

4. Discussion

This paper proposes an ELM classifier based on depth structure. Choosing the appropriate number of hidden nodes under the D-ELM structure provides enhanced stability and generalization ability in the network. To ensure accurate tests and prevent node redundancy, when the number of hidden nodes was 100, the test accuracy of D-ELM was maintained at a relatively stable value over repeated tests. The accuracy was phased as shown in Figure 10. D-ELM significantly outperformed ELM with small fluctuations in amplitude, robustness to the number of test iterations, and higher network stability.

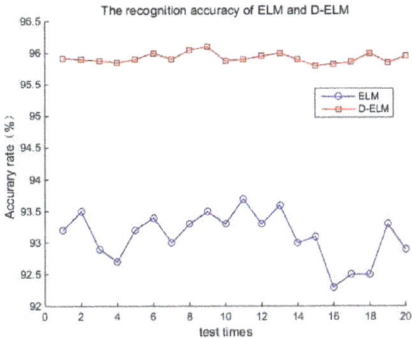

Figure 10. Accuracy of D-ELM and ELM after multiple tests.

Table 3 shows the results of our algorithm accuracy tests, as mentioned above. D-ELM has a higher average accuracy rate but lower standard deviation than ELM. The accuracy and stability of D-ELM network were both accurate and stable. As a result, the performance of the classifier was improved.

Table 3. D-ELM versus ELM stability.

Algorithms	Average Accuracy Rate (%)	Highest Accuracy Rate (%)	Lowest Accuracy Rate (%)	Standard Deviation (std)
D-ELM	95.92	96.11	95.65	0.0967
ELM	92.84	93.72	92.16	0.3220

We added an SVM classifier to the experiment to further assess the depth extreme learning machine. Table 4 shows the accuracy and timing of D-ELM and SVM training tests on all samples, where D-ELM again has the highest accuracy in both training and testing. Although the training time and network layer quantity are higher in D-ELM, its training time and test time are shorter than the other algorithms we tested, and its accuracy is much higher. The overall performance of D-ELM is better than that of ELM and SVM.

Table 4. Defect recognition of D-ELM ELM and SVM on wood images.

	Algorithm	D-ELM	ELM	SVM
Train	Time (s)	890	1879	2356
	Accuracy rate (%)	99.47%	92.31%	90.45%
Test	Time (ms)	187	532	631
	Accuracy rate (%)	96.72	92.16	91.55

5. System Interface

We constructed a network model and classification optimizer based on the proposed algorithm by integrating Anaconda 3.5 and TensorFlow. We then constructed a real wood plate defect identification system in the C# development language on the Microsoft Visual Studio 2017 open platform. The system can identify defects in solid wood plate images and provide their position and size information. We also used a Microsoft SQL server 2012 database to store the information before and after processing.

Our experiment on deep network feature learning mainly involved the implementation of the network framework and the training model for wood recognition. The system is based on the network training model discussed in this paper; it can be used to detect defects in the wood image database on a single sheet and display the coordinates in the X and Y directions of the defects, as shown in Figure 11. On the left side, the scanned wood images are displayed with defects marked in green boxes. The top right side of the interface shows the coordinates of each defect, the cutting position of the plank, and the type of defects. Below the table are the total numbers of defects identified by the machine and the recognition rate.

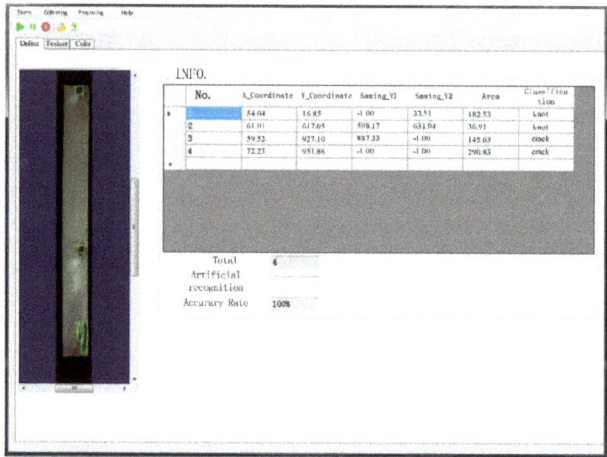

Figure 11. System interface diagram based on depth learning.

6. Conclusions

The depth extreme learning machine proposed in this paper has reasonable dimensions, effectively manages heterogeneous data, and works within an acceptable run time. Our results suggest that it is a promising new solution to problems such as obtaining marking samples, constructing features, and training. It has excellent feature extraction ability and fast training time. Based on the method of machine learning, The NSST transform is used to preprocess the original image (i.e., reduce its complexity and dimensionality while minimizing the down-sampling process in CNN), then SLIC is applied to optimize the CNN model training process. This method effectively reduces the redundancy of local image information and extracts relatively accurate supplementary feature contours. The optimized CNN is then used to extract wood image features and secure corresponding image features. The feature is input to the ELM classifier and the parameters of the related neural network are optimized. The GA is used to select the initial weight threshold of ELM to improve the prediction accuracy and stability of the network model. Finally, the image data to be tested is input to the well-trained network model and final test results are obtained.

We also compared the stability of D-ELM and ELM network models. The standard deviation of D-ELM was only 0.0967 and the accuracy of D-ELM improved by about 3% compared to ELM; the stability of the D-ELM network was also found to be higher and less affected by test quantity than ELM. We also found that D-ELM has an accuracy of up to 96.72% and a shorter test time than ELM or SVM at only 187 ms. The D-ELM network model is capable of highly accurate wood defect recognition within a very short training and detection time.

Author Contributions: Conceptualization, Y.Y. and Y.L.; methodology, Y.Y.; software, X.Z.; validation, Y.Y., Z.H. and F.D.; resources, X.Z. and Z.H.; writing—original draft preparation, Y.Y.; writing—review and editing, Y.L. and X.Z.; project administration, Y.L.; funding acquisition, Y.L. All authors have read and agreed to the published version of the manuscript.

Funding: This research was funded by the 2019 Jiangsu Province Key Research and Development Plan by the Jiangsu Province Science and Technology under grant BE2019112, and was funded by the Jiangsu Province International Science and Technology Cooperation Project under grant BZ2016028, and was supported from the 948 Import Program on the Internationally Advanced Forestry Science and Technology by the State Forestry Bureau under grant 2014-4-48.

Conflicts of Interest: The authors declare no conflict of interest.

References

1. Qiu, Q.W.; Lau, D. Grain Effect on the Accuracy of Defect Detection in Wood Structure by Using Acoustic-Laser Technique. In Proceedings of the Nondestructive Characterization and Monitoring of Advanced Materials, Aerospace, Civil Infrastructure, and Transportation XIII, Denver, CO, USA, 3–7 March 2019. [CrossRef]
2. Siekanski, P.; Magda, K.; Malowany, K.; Rutkiewicz, J.; Styk, A.; Krzeslowski, J.; Kowaluk, T.; Zagorski, A. On-line laser triangulation scanner for wood logs surface geometry measurement. *Sensors* **2019**, *19*, 1074. [CrossRef] [PubMed]
3. Espinosa, L.; Prieto, F.; Brancheriau, L.; Lasaygues, P. Effect of wood anisotropy in ultrasonic wave propagation: A ray-tracing approach. *Ultrasonics* **2019**, *91*, 242–251. [CrossRef] [PubMed]
4. Taskhiri, M.S.; Hafezi, M.H.; Harle, R.; Williams, D.; Kundu, T.; Turner, P. Ultrasonic and thermal testing to non-destructively identify internal defects in plantation eucalypts. *Comput. Electron. Agric.* **2020**, *173*. [CrossRef]
5. Yang, H.M.; Yu, L. Feature extraction of wood-hole defects using wavelet-based ultrasonic testing. *J. For. Res.* **2017**, *28*, 395–402. [CrossRef]
6. Lukomski, M.; Strojecki, M.; Pretzel, B.; Blades, N.; Beltran, V.L.; Freeman, A. Acoustic emission monitoring of micro-damage in wooden art objects to assess climate management strategies. *Insight* **2017**, *59*, 256–264. [CrossRef]
7. Rescalvo, F.J.; Valverde-Palacios, I.; Suarez, E.; Roldan, A.; Gallego, A. Monitoring of carbon fiber-reinforced old timber beams via strain and multiresonant acoustic emission sensors. *Sensors* **2018**, *18*, 1224. [CrossRef] [PubMed]
8. Li, C.; Zhang, Y.; Tu, W.; Jun, C.; Liang, H.; Yu, H. Soft measurement of wood defects based on LDA feature fusion and compressed sensor images. *J. For. Res.* **2017**, *28*, 1285–1292. [CrossRef]
9. Dai, J.; Li, Y.; He, K.; Sun, J. R-FCN: Object detection via region-based fully convolutional networks. *arXiv* **2016**, arXiv:1605.06409.
10. Szegedy, C.; Liu, W.; Jia, Y.Q.; Sermanet, P.; Reed, S.; Anguelov, D.; Erhan, D.; Vanhoucke, V.; Rabinovich, A. Going Deeper with Convolutions. In Proceedings of the 2015 IEEE Conference on Computer Vision and Pattern Recognition (CVPR), Boston, MA, USA, 7–12 June 2015; pp. 1–9. [CrossRef]
11. LeCun, Y.; Bengio, Y.; Hinton, G. Deep learning. *Nature* **2015**, *521*, 436–444. [CrossRef]
12. He, T.; Liu, Y.; Yu, Y.B.; Zhao, Q.; Hu, Z.K. Application of deep convolutional neural network on feature extraction and detection of wood defects. *Measurement* **2020**, *152*. [CrossRef]
13. Hu, K.; Wang, B.J.; Shen, Y.; Guan, J.R.; Cai, Y. Defect identification method for poplar veneer based on progressive growing generated adversarial network and MASK R-CNN Model. *Bioresources* **2020**, *15*, 3041–3052. [CrossRef]
14. Shi, J.H.; Li, Z.Y.; Zhu, T.T.; Wang, D.Y.; Ni, C. Defect detection of industry wood veneer based on NAS and multi-channel mask R-CNN. *Sensors* **2020**, *20*, 4398. [CrossRef] [PubMed]
15. Yang, W.X.; Jin, L.W.; Tao, D.C.; Xie, Z.C.; Feng, Z.Y. DropSample: A new training method to enhance deep convolutional neural networks for large-scale unconstrained handwritten Chinese character recognition. *Pattern Recogn.* **2016**, *58*, 190–203. [CrossRef]
16. Wan, W.; Lee, H.J. Multi-focus image fusion based on non-subsampled shearlet transform and sparse representation. *Lect. Notes Electr. Eng.* **2018**, *449*, 120–126. [CrossRef]
17. Singh, S.; Anand, R.S.; Gupta, D. CT and MR image information fusion scheme using a cascaded framework in ripplet and NSST domain. *IET Image Process* **2018**, *12*, 696–707. [CrossRef]
18. Wu, W.; Qiu, Z.M.; Zhao, M.; Huang, Q.H.; Lei, Y. Visible and infrared image fusion using NSST and deep Boltzmann machine. *Optik* **2018**, *157*, 334–342. [CrossRef]
19. Boemer, F.; Ratner, E.; Lendasse, A. Parameter-free image segmentation with SLIC. *Neurocomputing* **2018**, *277*, 228–236. [CrossRef]
20. Chen, L.C.; Papandreou, G.; Kokkinos, I.; Murphy, K.; Yuille, A.L. DeepLab: Semantic image segmentation with deep convolutional nets, atrous convolution, and fully connected CRFs. *IEEE Trans. Pattern Anal. Mach. Intell.* **2018**, *40*, 834–848. [CrossRef]

21. Zhu, W.T.; Miao, J.; Qing, L.Y.; Huang, G.B. Hierarchical Extreme Learning Machine for Unsupervised Representation Learning. In Proceedings of the Joint Conference on Neural Networks, Killarney, Ireland, 11–16 July 2015.
22. Uzair, M.; Shafait, F.; Ghanem, B.; Mian, A. Representation learning with deep extreme learning machines for efficient image set classification. *Neural Comput. Appl.* **2018**, *30*, 1211–1223. [CrossRef]
23. Tang, J.X.; Deng, C.W.; Huang, G.B. Extreme learning machine for multilayer perceptron. *IEEE Trans. Neural Networks Learn. Syst.* **2016**, *27*, 809–821. [CrossRef]
24. Yu, W.C.; Zhuang, F.Z.; He, Q.; Shi, Z.Z. Learning deep representations via extreme learning machines. *Neurocomputing* **2015**, *149*, 308–315. [CrossRef]
25. Liu, S.L.; Feng, L.; Xiao, Y.; Wang, H.B. Robust activation function and its application: Semi-supervised kernel extreme learning method. *Neurocomputing* **2014**, *144*, 318–328. [CrossRef]

Publisher's Note: MDPI stays neutral with regard to jurisdictional claims in published maps and institutional affiliations.

© 2020 by the authors. Licensee MDPI, Basel, Switzerland. This article is an open access article distributed under the terms and conditions of the Creative Commons Attribution (CC BY) license (http://creativecommons.org/licenses/by/4.0/).

Article

A Division Algorithm in a Redundant Residue Number System Using Fractions

Nikolay Chervyakov [1], Pavel Lyakhov [1,2,*], Mikhail Babenko [1], Irina Lavrinenko [3], Maxim Deryabin [1], Anton Lavrinenko [3], Anton Nazarov [1], Maria Valueva [1], Alexander Voznesensky [2] and Dmitry Kaplun [2]

1 Department of Applied Mathematics and Mathematical Modeling, Institute of Mathematics and Natural Sciences, North-Caucasus Federal University, 355000 Stavropol, Russia; ncherviakov@ncfu.ru (N.C.); mgbabenko@ncfu.ru (M.B.); maderiabin@ncfu.ru (M.D.); kapitoshking@mail.ru (A.N.); mriya.valueva@mail.ru (M.V.)
2 Department of Automation and Control Processes, Saint Petersburg Electrotechnical University "LETI", 197376 Saint Petersburg, Russia; a-voznesensky@yandex.ru (A.V); dikaplun@etu.ru (D.K.)
3 Department of Higher Algebra and Geometry, Institute of Mathematics and Natural Sciences, North-Caucasus Federal University, 355000 Stavropol, Russia; lavrinenko_ir1@mail.ru (I.L.); k-fmf-primath@stavsu.ru (A.L.)
* Correspondence: ljahov@mail.ru

Received: 3 December 2019; Accepted: 15 January 2020; Published: 19 January 2020

Abstract: The residue number system (RNS) is widely used for data processing. However, division in the RNS is a rather complicated arithmetic operation, since it requires expensive and complex operators at each iteration, which requires a lot of hardware and time. In this paper, we propose a new modular division algorithm based on the Chinese remainder theorem (CRT) with fractional numbers, which allows using only one shift operation by one digit and subtraction in each iteration of the RNS division. The proposed approach makes it possible to replace such expensive operations as reverse conversion based on CRT, mixed radix conversion, and base extension by subtraction. Besides, we optimized the operation of determining the most significant bit of divider with a single shift operation of the modular divider. The proposed enhancements make the algorithm simpler and faster in comparison with currently known algorithms. The experimental simulation using Kintex-7 showed that the proposed method is up to 7.6 times faster than the CRT-based approach and is up to 10.1 times faster than the mixed radix conversion approach.

Keywords: residue number system; redundant residue number system; modular division; fraction; algorithm

1. Introduction

The residue number system (RNS) has attracted many researchers as a basis for computing, and the interest taken in it has increased dramatically over the latest decade, which could be seen from the large number of papers focusing on the practical application of RNS in digital signal processing, image processing systems, cryptographic systems, quantum automated machines, neural computers systems, massive concurrency of operations, cloud computing, etc. [1–7].

RNS, if compared to other scales of notation, offers the advantage of rapid addition and multiplication, which causes stirs of interest in the RNS in areas requiring large amounts of computation. However, some operations, such as comparison and division of numbers, are very complicated in the RNS. Finding faster division algorithms would allow detecting more promising new areas to apply RNS.

The known RNS division algorithms [8–23] can be divided into two classes: based on the comparison of numbers, and based on the subtraction.

The algorithm for integer division operates similarly to a conventional binary division proposed in [1,2]. This algorithm and its modifications have a major drawback, namely that each iteration requires a comparison of numbers.

The algorithm without these drawbacks, as proposed in [1,2], is based on replacing the divider by an approximate value, which may be the product of one or several RNS modules. The algorithm provides a correct result for the condition $b \leq \bar{b} < 2b$, where b is an actual divider and $\bar{b}$ is an approximation of b. It is easy to see that this condition cannot be satisfied for all moduli sets (for example: $p_1 = 9, p_2 = 11$, $b = 4$).

The main disadvantages of this algorithm are the necessity of mixed radix conversion (MRC) and scaling operations use, and special logic and tables for determining the approximate divider. There have been proposed several algorithms for solving the problem of division based on a comparison of numbers and methods of determining the sign, which can be classified as follows: [8,10,15] using MRC, [9] to formulate the problem in terms of determining the even numbers, and [11] using the base extension operation in iterations. All the proposed algorithms, however, have the disadvantage of long computation time and high hardware costs due to the use of MRC, Chinese remainder theorem (CRT), and other costly operations.

In [12–14,16], a high-speed division algorithm is presented, which uses the comparison of higher degrees of dividend and divisor instead of using the MRC and CRT for the division of modular numbers. The time complexity and hardware costs in these algorithms are smaller than other algorithms, although this algorithm contains redundant stages. To speed up the calculation of the current quotients, Hung and Parhami suggested a division algorithm based on parity checking, in which the quotient calculation occurs two times faster than the algorithms [14,16]. However, the calculation of the higher powers of two is time-consuming in the RNS, which are carried out in each cycle.

The known algorithm of division in the RNS format, in addition to the RNS moduli set, also uses a replacement module system, which is an auxiliary to preserve the dividend and divisor residues. Presented in the RNS dividend and divisor are converted into a variety of RNS presentations with the various modules of the system [18]. Using two moduli sets of RNS leads to a large redundancy and the necessity for direct and reverse conversion from moduli set to the auxiliary and back for the division operation, which drastically reduces its speed. A fast algorithm for the division based on the use of the index over the Galois field transformation $GF(p)$ is proposed in [18], which was simply implemented using LUTs (Look-Up Table). However, this algorithm is effective when processing data no more than 6–10 bits and when a modulus is a prime number. Thus, this algorithm is not efficient for large RNS ranges.

Most of the known iterative algorithms contain a large number of operations in each iteration. According to the authors, the algorithm based on the CRT with fractions considered in [11,17,23] is the best and has the time complexity $O(nb)$, where n is the number of RNS modules and b is the number of bits in each module, assuming that the value of each module is more or less the same. The disadvantage of this algorithm is a set of operations performed in each iteration: the operations of addition, multiplication, comparison, and parity checking. Furthermore, the execution of the algorithm requires the conversion of the quotient from the system {−1, 0, 1} into the system {0, 1}, which gives an additional burden on the runtime of a modular division of number procedures.

In this paper, we propose an algorithm for division in the general case in the RNS using only the register shifts and summations. The improved algorithm has the following properties: it is very fast compared to the algorithms that are still available; has no restrictions on the dividend and divisor (except for when the divisor is equal to zero); it does not use a preliminary estimate of the coefficient; it does not use the back divisor; and does not use the base expansion operation. In [20], a similar approach based on mixed radix number system (MRNS) is proposed; however, in addition to the original RNS moduli set, it also used an auxiliary modulus set, which requires additional calculations for data conversion and significantly slows down the calculation of the division result. The proposed algorithm allows increasing the performance of the division algorithm by using the CRTf method. In [14,16] the idea of the most significant bits for a quotient was proposed for an RNS with special moduli sets $\{2^k, 2^k - 1, 2^{k-1} - 1\}$ and $\{2^k + 1, 2^k, 2^k - 1\}$, while in the proposed work, this approach is expanded to the case of general moduli set.

The main difference between this paper and [22] is that in this paper, a division algorithm for redundant RNS is proposed. Redundant RNS is intended for the organization of fault-tolerant calculations, while its modules are separated into informational, by which information is encoded, and redundant, necessary to restore information in case of errors. Separation of modules into information and redundant allows simplifying calculations by taking into account the information and redundant range of the system.

The known algorithms for dividing the numbers represented in the RNS are based on the absolute values of the dividend and the divisor. In this paper, we do not use the absolute values but their relative values, which allows reducing the computational complexity of division algorithms.

The rest of the paper is organized as follows: Section 2 describes the basics of RNS (Section 2.1) and approximate method for determining the placement of the number in it (Section 2.2). The proposed RNS division algorithm is presented in Section 2.3. Results and discussion are presented in Section 3.

2. An Approximate Method for Determining the Positional Feature of the Modular Number

2.1. Residue Number System

In the RNS, a positive integer is represented as a bank of residues to selected co-prime bases. This approach allows one to replace operations with large numbers by operations with small numbers, which are represented as residues of the division of large numbers by earlier selected relatively prime modules $p_1, p_2, \ldots, p_n$. Let

$$A \equiv \alpha_1 (\bmod p_1), \ A \equiv \alpha_2 (\bmod p_2), \ \ldots, \ A \equiv \alpha_n (\bmod p_n). \tag{1}$$

Then, an integer A can be associated with the set $(\alpha_1, \alpha_2, \ldots, \alpha_n)$ of the least non-negative residues over one of the corresponding numbers. This correspondence will be one-to-one until $A < p_1 p_2 \ldots p_n$, according to the CRT. The set $(\alpha_1, \alpha_2, \ldots, \alpha_n)$ can be considered as one of the methods of the representation of the integer A in a computer, i.e., the modular representation or representation in the RNS.

The main advantage of this representation is the fact that the addition, subtraction, and multiplication operations are implemented very simply by the formulas:

$$A \pm B = (\alpha_1, \alpha_2, \ldots, \alpha_n) \pm (\beta_1, \beta_2, \ldots, \beta_n) = ((\alpha_1 \pm \beta_1) \bmod p_1, (\alpha_2 \pm \beta_2) \bmod p_2, \ldots, (\alpha_n \pm \beta_n) \bmod p_n) \tag{2}$$

$$A \times B = (\alpha_1, \alpha_2, \ldots, \alpha_n) \times (\beta_1, \beta_2, \ldots, \beta_n) = ((\alpha_1 \times \beta_1) \bmod p_1, (\alpha_2 \times \beta_2) \bmod p_2, \ldots, (\alpha_n \times \beta_n) \bmod p_n) \tag{3}$$

These operations are called modular, since, for their execution in the RNS, it is sufficient to fulfill one cycle of processing numerical values. In addition, this processing occurs in parallel, and the information value in each modulo channel does not depend on the other modulo channels.

Thus, there are three main advantages of RNS [1].

1. There is no carry propagation between RNS arithmetic units. Large numbers represented in the form of small residues that leads to faster data processing.
2. When using the RNS, large numbers are encoded into a set of small residues, which reduces the complexity of the arithmetic units and simplifies the computing system.

3. RNS is a non-positional system with independent arithmetic units; therefore, an error in one channel does not apply to others. Thus, the processes of error detection and error correction are simplified.

However, such operations as sign detection, comparison, division, and some others are time-consuming and expensive in the RNS [4].

2.2. Approximate Method

An analysis of difficult (non-modular) operations has shown that they can be represented exactly or approximately, so the methods for calculating positional characteristics can be divided into two groups:

– Methods for accurate calculation of positional characteristics.
– Methods for the approximate calculation of positional characteristics.

The methods for accurate calculation of positional characteristics are discussed in [1–3]. In this paper, we investigate the approximate method for calculating positional characteristics that can significantly reduce the hardware and time costs due to operations performed on positional codes of reduced capacity. In this regard, there is an issue of using the approximate method when calculating a certain number of non-modular procedures: determining intervals of numbers; number sign; number comparison, in cases where there is no need to know the exact value; and the difference between the numbers.

The point of the approximate method for calculating the positional characteristics of modular number is based on employing the relative values of the analyzed numbers to the full range defined by the CRT, which connects the positional number a with its representation in the remainder $(\alpha_1, \alpha_2, \ldots, \alpha_n)$, where α_i is the smallest non-negative residues of the number in relation to the modules of the residue number system $p_1, p_2, \ldots, p_n$ with the following expression:

$$a = \left| \sum_{i=1}^{n} \frac{P}{p_i} |P_i^{-1}|_{p_i} \alpha_i \right|_P \qquad (4)$$

where p_i are RNS modules, $P = \prod_{i=1}^{n} p_i$ is the range of RNS, $P_i = \frac{P}{p_i} = p_1 p_2 \ldots p_{i-1} p_{i+1} \ldots p_n$, and $|P_i^{-1}|_{p_i}$ is a multiplicative inversion of P_i modulo p_i.

If we divide the left and right parts of Expression (4) by the constant P, corresponding to the range of numbers, we will get the approximate value

$$\left| \frac{a}{P} \right|_1 = \left| \sum_{i=1}^{n} \frac{|P_i^{-1}|_{p_i}}{p_i} \alpha_i \right|_1 \approx \left| \sum_{i=1}^{n} k_i \alpha_i \right|_1, \qquad (5)$$

where $|*|_1$ denotes the fraction of $*$ (or Modulo 1 operation) [24], $k_i = \frac{|P_i^{-1}|_{p_i}}{p_i}$ are constants of the chosen system, and α_i are positions of the number represented in the RNS in modules p_i, where $i = 1, 2, \ldots, n$, and the value of the Expression (5) will be in the range $[0, 1)$. The result of the sum shall be found after summation and discarding the integer part while maintaining the fractional part of the sum. The fractional value $F(a) = \left| \frac{a}{P} \right|_1 \in [0, 1)$ contains information both on the magnitude of the number and on its sign. If $\left| \frac{a}{P} \right|_1 \in [0, \frac{1}{2})$, then the number a is positive and $F(a)$ is equal to the number of a, divided by P. Otherwise, a is a negative number, and $1 - F(a)$ indicates a relative value. Rounding $F(a)$ to 2^{-t} bits will be denoted as $[F(a)]_{2^{-t}}$. The exact value of $F(a)$ is determined by inequalities $[F(a)]_{2^{-t}} < F(a) < [F(a)]_{2^{-t}} + 2^{-t}$. The integer part, obtained through summing the constants k_i, is a rank number; that is, a non-positional feature that shows how many times the range of the system P was surpassed while passing from the number representation in the residue number system to its positional representation. If necessary, the rank can be determined directly through the operation of the summation the constants k_i. The fractional part can also be written as $A \bmod 1$ because $A = \lfloor A \rfloor + A \bmod 1$. The number of positions in the fractional part of the number is determined by the maximum potential difference between the adjacent numbers. In case of accurate comparison, which is widely used in the division of numbers, you need to calculate a value that is equivalent to the conversion of the RNS into the positional notation.

Rounding the $F(a)$ value will inevitably result in an error. Let us denote $\rho = -n + \sum_{i=1}^{n} p_i$. Work [22] shows that it is necessary to use $N = \lceil \log_2(P\rho) \rceil$ bits after the decimal point when rounding the value $F(a)$, so that the resulting error has no effect on the accuracy of calculations. In other words, there is established a one-to-one correspondence between the set of numbers represented in the RNS and the plurality of $[F(a)]_{2^{-N}}$ values. Using the variables $[F(a)]_{2^{-N}}$ in calculations, in terms of algorithmic complexity, is equivalent to applying the inverse transformation from the RNS into the positional notation using the CRT. This method is slow and therefore, in practice, the use of calculations with the values $[F(a)]_{2^{-N}}$ is not rational. In [22], it is shown that it is possible to use the values $[F(a)]_{2^{-\widetilde{N}}}$, where $\widetilde{N} < N$, for operations of determining the number sign in the RNS. The point of this approach is based on the fact that when determining the sign there is no need to know the exact value of the number, and it is just enough to know about the range within which the number tested falls.

The algorithm for determining the sign of the number serves the basis for number comparison algorithms. Determining the sign of the number in the RNS using the values $[F(a)]_{2^{-t}}$, takes the following operations:

1. «Rough estimate». At this stage, the value $[F(a)]_{2^{-\widetilde{N}}}$, $\widetilde{N} < N$ is used. If $0 < [F(a)]_{2^{-\widetilde{N}}} < \frac{1}{2} - 2^{-\widetilde{N}}\rho$, then the number a is positive. If $\frac{1}{2} < [F(a)]_{2^{-\widetilde{N}}} < 1 - 2^{-\widetilde{N}}\rho$, then the number a is negative.

2. «Clarification». If the number a falls into none of the intervals as indicated in Step 1, then a rechecking of the number is needed with maximum precision calculation regarding the $[F(a)]_{2^{-N}}$ values. If $0 < [F(a)]_{2^{-N}} < \frac{1}{2}$, then the number a is positive. If $\frac{1}{2} < [F(a)]_{2^{-N}} < 1$, then the number a is negative.

The speed of the algorithm at the stage of the «rough estimate» depends on how little the value $\widetilde{N}$ is compared to N. However, if $\widetilde{N}$ is taken as too little, then the intervals in Step 1 may be so small that the algorithm for numerous numbers in the RNS would require the use of the «clarification» stage, while the benefit of using a small capacity at the «rough estimate» stage would be dismissed completely. For example, in [13], it is proposed to use the case when $\widetilde{N} = 4$ that is usually too small for practice. Instead, we suggest using an estimation from [22], which shows that the optimal speed of the algorithm is achieved with $\widetilde{N} \approx \log_2(N\rho \ln 4)$. Here below comes a comparison of the $\widetilde{N}$ and N capacities for the RNS, where the ranges of 16, 32, and 64 bits are implemented.

1. Sixteen bits. RNS modules 7, 17, 19, 29.

$$P = 65569;\ \rho = 68;\ N = 23;\ \widetilde{N} = 11.$$

2. Thirty-two bits. RNS modules 2, 3, 5, 11, 13, 19, 23, 29, 79.

$$P = 4295006430;\ \rho = 175;\ N = 40;\ \widetilde{N} = 13.$$

3. Sixty-four bits. RNS modules 2, 11, 17, 19, 23, 31, 41, 53, 59, 61, 71, 79, 83.

$$P = 18446748995286100082;\ \rho = 537;\ N = 74;\ \widetilde{N} = 16.$$

Thus, with an RNS of a 16-bit range, the «rough estimate» is done using the values with a capacity of $\widetilde{N} = 11$ bits, while the «clarification» takes place at the $N = 23$ bit precision. The speed of the rough estimate goes up by 2.09 times. For an RNS with a 32-bit range, the «rough estimate» employs a capacity of $\widetilde{N} = 13$ bits, while the «clarification» requires $N = 40$ bits for calculation. The speed of the rough estimate increases 3.08 times. For a 64-bit RNS, the «rough estimate» would use a capacity of $\widetilde{N} = 16$ bits, while the «clarification» requires $N = 74$ bits. The speed of the rough estimate increases 4.62 times. These results show that, for large ranges, the capacity $\widetilde{N}$ employed for the rough estimate is significantly lower than the accurate calculation capacity N, which allows significant gains in terms of speed when performing non-modular operations.

Figure 1 shows the location of the mentioned intervals for positive and negative numbers in the RNS, and the location of the ambiguity areas, where it is possible to wrongly determine the sign. For the redundant RNS, the numerical range shows a redundancy zone. This allows reducing the number of the checked conditions due to the fact that the sets of the admissible positive numbers and the areas of the erroneous sign determination would no longer intersect (Figure 2). Thus, when speaking of a redundant RNS, determining the sign is reduced to the following tasks.

1. «Rough estimate». If $0 < [F(a)]_{2^{-\widetilde{N}}} < \frac{1}{2}$, then the number a is positive. If $\frac{1}{2} < [F(a)]_{2^{-\widetilde{N}}} < 1 - 2^{-\widetilde{N}}\rho$, then the number a is negative.
2. «Clarification». If the number a is not included in any of the intervals in Step 1, then there is a sign rechecking needed using the $[F(a)]_{2^{-N}}$ variables. If $0 < [F(a)]_{2^{-N}} < \frac{1}{2}$, then the number a is positive. If $\frac{1}{2} < [F(a)]_{2^{-N}} < 1$, then the number a is negative.

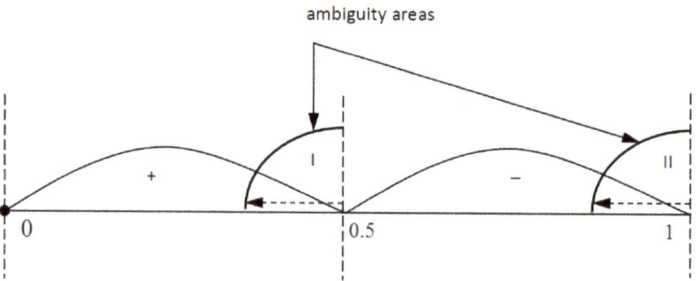

Figure 1. Position of ambiguity areas when determining the sign and the intervals for positive and negative numbers for irredundant residue number system (RNS).

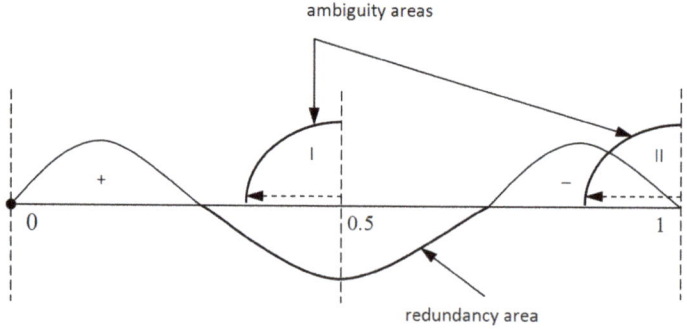

Figure 2. Position of ambiguity areas when determining the sign and the intervals for positive and negative numbers for redundant RNS.

Let us have a view on employing the approximate method by comparing the numbers in the RNS.

Example 1. *We have a system of bases $p_1 = 2$, $p_2 = 3$, $p_3 = 5$, $p_4 = 7$.*

Then $P = 2 \cdot 3 \cdot 5 \cdot 7 = 210$, $\rho = 2 + 3 + 5 + 7 - 4 = 13$, $P_1 = \frac{P}{p_1} = 105$, $P_2 = \frac{P}{p_2} = 70$, $P_3 = \frac{P}{p_3} = 42$, $P_4 = \frac{P}{p_4} = 30$.

The constants k_i used for computing the relative values are:

$$k_1 = \frac{\left|\frac{1}{105}\right|_2}{2} = \frac{1}{2}; \; k_2 = \frac{\left|\frac{1}{70}\right|_3}{3} = \frac{1}{3}; \; k_3 = \frac{\left|\frac{1}{42}\right|_5}{5} = \frac{3}{5}; \; k_4 = \frac{\left|\frac{1}{30}\right|_7}{7} = \frac{4}{7}.$$

For precise operations with relative sizes of numbers in the RNS, it is necessary to use $N = \lceil \log_2(P\rho) \rceil = 12$ characters after the decimal point. For a quick «rough estimate», we will use $\widetilde{N} = \lfloor \log_2(N\rho \ln 4) \rfloor = 7$ decimals. The constants k_i rounded up to 7 and 12 bits after the decimal point, are, respectively:

Seven bits: $k_1 = 0.1000000$; $k_2 = 0.0101010$; $k_3 = 0.1001100$; $k_4 = 0.1001001$;

Twelve bits: $k_1 = 0.100000000000$; $k_2 = 0.010101010101$; $k_3 = 0.100110011001$; $k_4 = 0.100100100100$.

The «rough estimate» takes checking the conditions $0 < [F(a\;)]_{2^{-7}} < \frac{1}{2} - 2^{-7}\rho$ and $\frac{1}{2} < [F(a\;)]_{2^{-7}} < 1 - 2^{-7}\rho$ (Step 1 of the algorithm), which in the binary form will appear as

1.1. If $0 < [F(a\;)]_{2^{-7}} < 0.0110011$, then the number a is positive.

1.2. If $0,1 < [F(a\;)]_{2^{-7}} < 0.1110011$, then the number a is negative.

Let us compare the two numbers $a = 97$ $b = 8$ presented in the RNS on the bases p_1, p_2, p_3, and p_4. Let us define the numbers a and b in the RNS as: $a = (1, 1, 2, 6)$, $b = (0, 2, 3, 1)$. The difference is $a - b = (1, 1, 2, 6) - (0, 2, 3, 1) = (1, 2, 4, 5)$. We will also define the sign $a - b$. For the «rough estimate», we will find that $[F(a - b)]_{2^{-7}} = 0.0110001$. The values found meets the condition of Step 1 of the algorithm, that is $0 < 0.0110001 < 0.0110011$, so we can conclude that $a - b > 0$, which produces $a > b$.

Now, let us compare the two numbers $a = 97$ and $b = 96$ as presented in the RNS on the bases p_1, p_2, p_3, and p_4. Now, we will define the numbers a and b in the RNS: $a = (1, 1, 2, 6)$, $b = (0, 0, 1, 5)$. The difference is $a - b = (1, 1, 2, 6) - (0, 0, 1, 5) = (1, 1, 1, 1)$. We will define the sign $a - b$. For the «rough estimate», $[F(a - b)]_{2^{-7}} = 0.1111111$. None of the conditions are met regarding the value obtained, so it will take a clarification stage of the algorithm. For the «accurate estimation», we will find $[F(a - b)]_{2^{-12}} = 0.000000010010$. This value follows the condition of Step 2 of the algorithm, so we conclude that $a - b > 0$, where $a > b$.

The example above serves an illustration of employing the approximate method for computing in the RNS. It has been shown how to take into account the error that occurs when using a small $\widetilde{N}$. In practice, for most cases, it would be enough to carry out a «rough estimate», a run wherein it takes operating with numbers whose capacity is close to the logarithm of the full range capacity. Therefore, the complexity of the «rough estimate» is committed to $O(\log_2 n)$, while the complexity of the «clarification» stage tends to $O(n)$.

2.3. Division Algorithm in the RNS

The algorithm for the $\frac{a}{b}$ integer division could be described with an iterative scheme, which is performed in two stages. The first stage implies a search for the highest power 2^i when approximating the quotient with a binary sequence. The second stage involves clarification of the approximating series. To get a range greater than P, you can select a value $P' = P \cdot p_{n+1}$; thus, it will take expanding the RNS base through adding an extra module. To avoid this base expansion, which is a computationally complex operation, we need to compare not the dividend with the interim divisors but the current results of the iteration (i) with the values of the previous iterations ($i - 1$). We will repeat the process of doubling the divider as long as the intermediate divider at the i iteration is below that of the $i - 1$ iteration. This would allow meeting the condition $0 < b < P - 1$.

The division algorithm can be described with the following rules.

A certain rule φ is constructed, which, for each pair of positive integers, a and b will assign a certain positive number q_i, where i is the number of the iteration, so that $a - bq_i = r_i > 0$, i.e., $a > bq_i$. Then, the division of a by b will follow the rule: based on the operation φ, each pair of a and b will be assigned a corresponding number $q_1 = q_0$, so that $a - bq_1 = r_1 \geq 0$, i.e., $a \geq bq_1$. We will take the values 2^i as q_i and place them into the memory as the constants $c_i = \left(2^i \bmod p_1, 2^i \bmod p_2, \ldots, 2^i \bmod p_n\right)$. Given that, the $i + 1$ operation does not depend on the i-th operation, which allows performing iterations in parallel. Furthermore, in each iteration, there are only two operations performed: multiplication of the constant divisor by 2^i, and comparison of the obtained values with the dividend.

If $r_1 \leq b$, then the division is complete; if $r_1 \geq b$, then following the rule φ, the pair of numbers (r_1, b) will get a q_2 assigned, so that $a - bq_2 = r_2 \geq 0$, i.e., $a \geq bq_2$. If $r_2 < b$, then the division is completed, and if $r_2 \geq b$, then following the rule φ, the pair of numbers (r_2, b) is assigned a q_3, so that $a - bq_3 = r_3 \geq 0$, etc. Since the consistent application of the operation φ leads to a decreasing sequence of integers $a > r_1 > r_2 > \ldots \geq 0$, then the algorithm is implemented in a finite number of steps. Let us assume that at step m there is a case $0 < bq_m$ recorded, which means the end of the division operation. Then, we finally obtain $a \cong (q_1 + q_2 + \ldots + q_m)b + r_m$, where the sequence $q_1 + q_2 + \ldots + q_m$ is the approximation of the quotient, which may contain some extra q_i. Next, we need clarification for the resulting approximating series. In [14] and [16], the idea of the most significant bits for the quotient was introduced for RNS with specialized moduli sets $\left\{2^k + 1, 2^k, 2^k - 1\right\}$ and $\{2^k, 2^k - 1, 2^{k-1} - 1\}$, while the approach proposed in this paper is extended for a general case.

The clarification will start with the higher q_m. If $a > bq_m$, then q_m is a member of the approximating series of the resulting quotient. Further, we take $(q_m + q_{m-1})$: if $a > b(q_m + q_{m-1})$, then q_{m-1} is put into the line, otherwise, if $a < b(q_m + q_{m-1})$, then q_m is excluded from the series, etc. After checking all the q_i, the quotient shall be determined by the remaining members of the series. Then, the quotient desired is determined by the expression $a = (q_m + q_{m-1} + \ldots + q_i + \ldots + q_1)b + r_m$, where

$$q_i = \begin{cases} 1, \text{ if } (q_m + q_{m-1} + \ldots + q_i)b < a; \\ 0, \text{ otherwise.} \end{cases}$$

This algorithm will be easy to modify it into a modular form, while the absolute values of the variables are replaced with their relative values. The structure of the algorithm proposed is based on employing the approximate method for comparing numbers, which is performed using subtraction.

The known algorithms determine the quotient on the basis of iteration $A' = A - QD$, where A and A', respectively, are the current and the next dividend, D is the divisor, Q_1 is the quotient, which is generated at each iteration of the full range of the RNS, and is not chosen from a small set of constants. In the proposed algorithm, the quotient is determined from the iteration $r_i = A - b2^i$, where A is a certain dividend, b—divisor, and 2^i is a member of the quotient's approximating series.

A comparison of the algorithms shows that the dividend in all iterations does not change, while the divisor is multiplied by the constant, which significantly reduces the computational complexity. In the iterative process of division in positional notation, in order to search for the highest power of the quotient's approximating series, and to clarify the approximating series, the dividend is compared to the doubled divisors or to the sum of the members of the series. Application of this principle to RNS can lead to incorrect operation of the algorithm, since, in case of the dynamic range overflow for the intermediate divider, the reconstructed number may go beyond the operating range caused by cyclic RNS. The cyclic RNS value will be below the dividend, which is not true because, in fact, the numbers will exceed the range P and the algorithm will proceed to the «loop» mode. For example, if the RNS modules are $p_1 = 2, p_2 = 3, p_3 = 5$, and $p_4 = 7$, then the range is $P = 2 \cdot 3 \cdot 5 \cdot 7 = 210$. Suppose the reconstruction produced the number $A = 220$. In the RNS, $A = 220 = (0, 1, 0, 3)$, i.e., $A = 210$ and $A' = 10$ have the same representation in the RNS. This ambiguity can lead to a breach of the algorithm. To overcome this difficulty, there is a need to compare the RNS the results of the current iteration values with the previous ones, which allows correct determination of a larger or smaller number. So, the fact of the dynamic range overflow in the RNS can be used for decision-making, «more-less». At the first iteration, there is a comparison of the dividend with the divisor, while the remaining iterations compare the doubled values of the divisors $q_i b < q_{i+1} b$. Each new iteration implies a comparison of the current value with the previous one.

Consistent application of these iterations leads to the formation of the inequalities chain $bq_1 < bq_2 < \ldots < bq_m > bq_{m+1}$, which determines the required number of iterations dependent on the values of the dividend and the divisor. Thus, the algorithm is implemented through a finite number of iterations. Suppose that at iteration $m+1$ there is a case of closure of the increasing sequence $bq_m > bq_{m+1}$, which corresponds to the RNS overflow range, i.e., $bq_{m+1} > P$ and $a < bq_{m+1}$. Here is the end of the process of developing quotient interpolation through a binary sequence or a set of constants in the RNS. Thus, the process of the quotient approximation can be done by comparing the neighboring approximate divisors.

Here below, we will provide a detailed description of an improved algorithm for the division of modular numbers in a redundant RNS.

2.4. Determination of the Quotient Sign

Step 1. Calculate the approximate values of the dividend $F(a)$ and the divisor $F(b)$. We determine the signs of the numbers in two stages.

1. «Rough estimate». If $0 < [F(a)]_{2^{-\tilde{N}}} < \frac{1}{2}$, then the number a is positive. If $\frac{1}{2} < [F(a)]_{2^{-\tilde{N}}} < 1 - 2^{-\tilde{N}} \rho$, then the number a is negative.
2. «Clarification». If the number a has not fallen into the range of Step 1, then a rechecking for the sign is needed, using the values $[F(a)]_{2^{-N}}$. If $0 < [F(a)]_{2^{-N}} < \frac{1}{2}$, then the number a is positive. If $\frac{1}{2} < [F(a)]_{2^{-N}} < 1$, then the number a is negative.

Step 2. If the numbers a and b have different signs, then the quotient is negative. If the numbers a and b have the same signs, then the quotient is positive. In further calculations, we use the absolute values of the divisor a and the divisor b. For the sake of convenience, we will denote them, too, as a and b.

Approximation of the Quotient

Step 3. Calculate the approximate values of the dividend $F(a)$ and the divisor $F(b)$ and compare them. If $F(a) \leq F(b)$, then the division process ends and the quotient $\lfloor \frac{a}{b} \rfloor$ is, respectively, equal to 0 or 1. If $F(a) > F(b)$, then there is a search for the highest power 2^{-N} in the approximation of the quotient with the binary code, where $-N$ is a least significant bit of the binary fraction.

Let us show the search for the highest degree in the binary fraction.

Step 4. Shift the function $[F(b)]_{2^{-N}}$ to the left up until a change in the first bit after the decimal point. The number of shifts determines the highest power j, which is recorded with the pulse counter connected to the memory V.

In this approximation, the quotient ends. To clarify the approximating sequence of the quotient, we will perform the following steps.

2.5. Clarification of the Quotient's Approximating Sequence

Step 5. From the memory, we select the constant 2^j (the highest power of the series) and multiply it by the divisor. The value $2^j F(b)$ will be compared with the dividend $F(a)$ using the approximate method of number comparison in the RNS.

The constants 2^j, $1 \leq j \leq \log_2 P$ are previously placed in the memory V; the counter j and the quotient Q are set on «0». The outputs of the counter are address inputs in the memory V.

Step 6. Calculate the $\Delta_1 = F(a) - F_1(b)$. If the sign bit Δ_1 the value is «1», then the corresponding power series is discarded; if the value is «0», then to the quotient adder we add the value of the sequence members with the same degree, i.e., $2^j \bmod p_i$, $1 \leq i \leq n$, $0 \leq j \leq N$.

Step 7. Check the sequence member of the 2^{j-1} degree through a shift to the right and comparison. Compare Δ_1 and $2^{j-1}b$. If $\Delta_1 < 2^{j-1}b$, then the corresponding power series is discarded; if $\Delta_1 > 2^{j-1}b$, then to the quotient adder we add the value of the sequence members with the same degree, i.e., $2^{j-1} \bmod p_i$ $\Delta_2 = \Delta_1 - 2^{j-1}b$.

Step 8. Similarly, check all the remaining sequence members up to degree zero. The last $\Delta_i = R = \Delta_{i-1} - F_{i-1}$, i.e., $0 \le R < b$ will be the remainder of a divided by b. The quotient Q will be the sum of all the 2^j needed for developing the quotient, which was accumulated in the adder with the sign as defined in the second step. The algorithm terminates.

The performance of the modified algorithm could be further shown with the example below.

Example 2. Find the quotient $Q = \frac{a}{b}$ of dividing $a = 97$ by $b = -8$ in an RNS with bases $p_1 = 2$, $p_2 = 3$, $p_3 = 5$, $p_4 = 7$. Then $P = 2 \cdot 3 \cdot 5 \cdot 7 = 210$, $\rho = 2 + 3 + 5 + 7 - 4 = 13$, $P_1 = \frac{P}{p_1} = 105$, $P_2 = \frac{P}{p_2} = 70$, $P_3 = \frac{P}{p_3} = 42$, and $P_4 = \frac{P}{p_4} = 30$.

The constants k_i used for calculation of the relative values are:

$$k_1 = \frac{\left|\frac{1}{105}\right|_2}{2} = \frac{1}{2};\ k_2 = \frac{\left|\frac{1}{70}\right|_3}{3} = \frac{1}{3};\ k_3 = \frac{\left|\frac{1}{42}\right|_5}{5} = \frac{3}{5};\ k_4 = \frac{\left|\frac{1}{30}\right|_7}{7} = \frac{4}{7}.$$

For a quick «rough estimate», we will use $\widetilde{N} = \lfloor \log_2(N\rho \ln 4) \rfloor = 7$ characters after the decimal point. The constants k_i rounded up to 7 bits after the decimal point are:

Seven bits: $k_1 = 0.1000000$; $k_2 = 0.0101010$; $k_3 = 0.1001100$; $k_4 = 0.1001001$.

Precise operations with relative values of the numbers in the RNS take $N = \lceil \log_2(P\rho) \rceil = 12$ characters after the decimal point. The constants k_i rounded up to 12 binary bits after the decimal point are:

$$k_1 = 0.100000000000;\ k_2 = 0.010101010101;\ k_3 = 0.100110011001;$$
$$k_4 = 0.100100100100.$$

Now, we shall represent the a b numbers in the RNS:

$$a_{10} = 97 \to (1, 1, 2, 6)_{RNS},$$

$$b_{10} = -8 \to (0, 1, 2, 6)_{RNS}.$$

Determine the signs of the numbers a and b.

A «rough estimate» (binary):

$$[F(a)]_{2^{-7}} = |1 \cdot 0.1000000 + 1 \cdot 0.0101010 + 10 \cdot 0.1001100 + 110 \cdot 0.1001001|_1 = 0.0111000.$$

Since $[F(a)]_{2^{-7}}$ misses any one of the intervals $(0; 0.0110011)$, $(0.1; 0.1110011)$ as set forth in Example 1 will take a clarifying iteration:

$$[F(b)]_{2^{-7}} = |0 \cdot 0.1000000 + 1 \cdot 0.0101010 + 10 \cdot 0.1001100 + 110 \cdot 0.1001001|_1 = 0.1111000$$

$[F(b)]_{2^{-7}}$, too, misses all the intervals $(0; 0.0110011)$, $(0.1; 0.1110011)$ as set forth in Example 1, so it will take another clarifying iteration.

«Clarification»:

$$[F(a)]_{2^{-12}} = |1 \cdot 0.100000000000 + 1 \cdot 0.010101010101 + 10 \cdot 0.100110011001 + 110 \cdot 0.100100100100|_1 =$$
$$= 0.011101011111$$

Since $0 < [F(a\,)]_{2^{-12}} < 0.1$, then the number a is positive.

$[F(b)]_{2^{-12}} = |0 \cdot 0.100000000000 + 1 \cdot 0.010101010101 + 10 \cdot 0.100110011001 + 110 \cdot 0.100100100100|_1 =$
$= 0.111101011111$

Since $0.1 < [F(b)]_{2^{-12}} < 1$, then the number b is negative.

The numbers a and have opposite signs, so the quotient sign will be negative. In order to find the absolute value of the quotient, we will divide a by $-b = (0, 0, 0, 0) - (0, 1, 2, 6) = (0, 2, 3, 1)$ following the algorithm as specified above.

The relative values of the dividend a and the divisor $-b$ with full accuracy of the calculations N are:

$$[F(a\,)]_{2^{-12}} = 0.011101011111;\ [F(-b)]_{2^{-12}} = 0.000010011001.$$

Shifting the fractional part of the divisor $-b$ to the left, step by step, we determine that a change in the first fractional bit after the decimal point occurs at the fourth shift. Thus, the approximation series can include only the values 2^0, 2^1, 2^2, and 2^3, which, in the RNS, have the following representation:

$$q_0 = 2^0 = (1, 1, 1, 1);\ q_1 = 2^1 = (0, 2, 2, 2);\ q_2 = 2^2 = (0, 1, 4, 4);\ q_3 = 2^3 = (0, 2, 3, 1).$$

These values develop the approximation sequence of the quotient, which is to be clarified later on. For a more accurate approximation sequence, we will subtract from the fraction $[F(a\,)]_{2^{-12}}$ of the dividend the fraction of the divisor $[F(-b)]_{2^{-12}}$ that has been shifted three ranks to the left (i.e., multiplied by 2^3):

$$\Delta_1 = [F(a\,)]_{2^{-12}} - 2^3 \cdot [F(-b)]_{2^{-12}} = 0.011101011111 - 0.010011001 = 0.001010010111.$$

Since $\Delta_1 > 0$, then we will leave 2^3 in the approximation sequence, while the value Δ_1 will be used for further calculations.

We subtract from Δ_1 the fraction $[F(-b)]_{2^{-12}}$ of the divisor shifted left two ranks:

$$\Delta_2 = \Delta_1 - 2^2 \cdot [F(-b)]_{2^{-12}} = 0.001010010111 - 0.0010011001 = 0.000000110011.$$

Since $\Delta_2 > 0$, then we leave 2^2 in the approximation sequence, while the value Δ_2 will be used for further calculations.

We subtract from Δ_2 the fraction $[F(-b)]_{2^{-12}}$ of the divisor shifted left one rank:

$$\Delta_3 = \Delta_2 - 2^1 \cdot [F(-b)]_{2^{-12}} = 0.000000110011 - 0.00010011001 = 1.111100000001$$

The appearance of 1 in the sign rank indicates that $\Delta_3 < 0$, therefore 2^1 is excluded from the approximation sequence, and Δ_3 is not to be used further (continue using Δ_2).

We subtract from Δ_2 the fraction $[F(-b)]_{2^{-12}}$ of the divisor (no shift applied):

$$\Delta_4 = \Delta_2 - 2^{01} \cdot [F(-b)]_{2^{-12}} = 0.000000110011 - 0.000010011001 = 1.111110011010.$$

The appearance of 1 in the sign rank indicates that $\Delta_4 < 0$, so 2^0 is excluded from the approximation sequence.

Here, the process of clarifying the approximation sequence comes to an end. To determine the quotient, we need to add the remaining members of the approximation sequence. In this example, the remaining members were the following ones: $q_3 = 2^3 = (0, 2, 3, 1)$ and $q_2 = 2^2 = (0, 1, 4, 4)$. Then the absolute value of the quotient is to be determined through summing the members of the sequence:

$$\left\| \frac{a}{b} \right\| = (0, 2, 3, 1) + (0, 1, 4, 4) = (0, 0, 2, 5) = 12.$$

In view of the sign, we finally obtain $\lfloor \frac{a}{b} \rfloor = -12$.

Figure 3 demonstrates the scheme of positional characteristics calculation based on CRTf for a number $X = \{x_1, x_2, \ldots, x_n\}$. A bit's width of values x_i is equal to $\lceil \log_2 p_i \rceil$, $i = 1, 2, \ldots, n$. The initial moduli $|x_i \cdot k_i|_{2^N}$, $i = 1, 2, \ldots, n$ generates partial products of constant multiplication. Then, they are summed by a Carry-Save-Adder-tree (CSA-tree) modulo 2^N. Obtained results are summed by Kogge–Stone adder [25] modulo 2^N and is equal to $[F(X)]_{2^{-N}}$. In the next section, we will demonstrate the advantages of the proposed method compared to known analogs based on CRT and MRC.

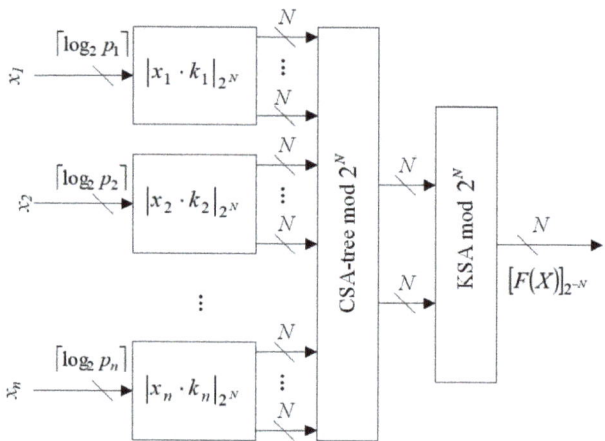

Figure 3. The scheme of positional characteristics calculation based on CRTf.

3. Simulation of the Proposed Algorithm

It follows from the analysis of the modular division scheme that the comparison and sign detection unit is the main component determining its computational complexity. This unit can be implemented based on the CRT, MRNS, or CRT with fractions. We have considered the models of all three types.

The experimental simulation has been performed using ISE Design Suite 14.7. Kintex-7 KC705 XC7K70T-2FBG676 without DSP48E1B blocks has been chosen as the goal of compilation. This FPGA contains 10,250 slices and 300 input–output blocks. During the simulation, we varied the digit capacity of the moduli under a fixed number of bases. For each type of the model, same prime bases of a given capacity have been selected; in particular, four bases with module bits 5, 9, 13, 17, 21, and 25. The dynamic range of the system is approximately the product of the number of bases and their digit capacities. Only the bottleneck of the RNS division algorithm was implemented in hardware. The remaining parts of the division algorithm are very similar to the division operation in the standard IEEE library "ieee.numeric_std.all" and require approximately the same amount of resources in hardware implementation. Figure 4 shows the resource usage graph of this FPGA with different capacity moduli for each type of the model. Table 1 shows detailed resource utilization for all approaches considered.

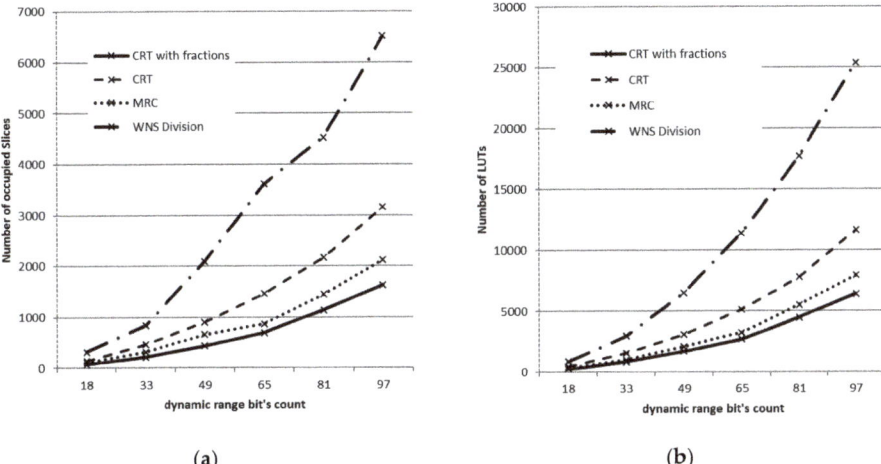

Figure 4. FPGA Kintex-7 KC705 XC7K70T-2FBG676 resource usage by selected calculation basis: (**a**) number of occupied Slices, (**b**) number of LUTs.

Table 1. Resources utilization and total delay.

Method	Utilization	Dynamic Range of RNS, Count of Bits					
		18	33	49	65	81	97
CRT with fractions	LUTs	253	843	1700	2684	4454	6367
	Slices	71	219	438	689	1135	1618
	Delay, ns	11.3	12.361	14.331	15.992	17.983	18.962
CRT	LUTs	473	1571	3064	5105	7743	11643
	Slices	138	467	896	1457	2165	3160
	Delay, ns	37.607	62.5	96.633	123.116	158.002	189.999
MRC	LUTs	325	1005	2073	3214	5507	7907
	Slices	113	321	662	865	1444	2117
	Delay, ns	37.412	70.032	117.226	161.908	196.296	285.723
WNS division	LUTs	865	2936	6474	11,395	17,703	25,389
	Slices	319	840	2093	3614	4526	6512
	Delay, ns	47.626	98.089	169.874	227.468	375.987	485.976

All the considered algorithms were implemented on the corresponding FPGA. The architectures for CRT and MRC from [4] were simulated for comparison. In addition, the results for the division algorithm in the weighted numeric system (WNS) are also presented. For division in the WNS, an algorithm from standard IEEE library "ieee.numeric_std.all" was implemented. The pipeline in the binary division is implemented like in [26]. These results clearly show the fact that the additional circuit logic for the proposed method using CRT with fractions does not exceed 25% of the WNS division algorithm costs.

We will comprehend the scheme latency as the maximum time spent by an arbitrary signal to run over the whole scheme from a certain input to a certain output. Latency estimation allows describing the performance of the suggested algorithm, including the working frequency of the scheme. For each type of model, Figure 5 presents the working frequency of the scheme for the base systems with different digit capacities of the moduli.

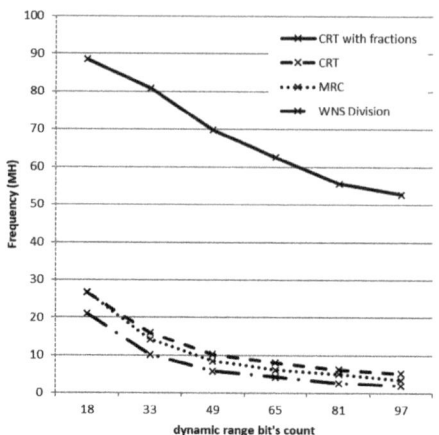

Figure 5. Frequency as a function of dynamic range bit count.

As an example, consider a 64-bit capacity as the most widespread in modern computer systems. To write numbers in this system, it suffices to represent each of the four moduli as a 16- or 17-bit number. Let the set of moduli be {65537, 65539, 65543, 65551}. The range of this set forms a 65-bit number, which covers 64-bit capacity. Here, the approximate method requires only 689 slices, whereas the orthogonal basis method and the improved MRNS scheme require 1457 and 865 slices, respectively. On the other hand, the working frequency of the approximate method reaches 62.5 MHz, which is 7.6 times faster than the CRT-based restoration and 10.1 times faster than the improved MRNS method. Note that the advantages of the approximate method over these approaches remain in force for higher digit capacities, too.

4. Conclusions

The new algorithm described in this paper speeds up the modular division procedure in the RNS representation in comparison with the well-known analogs. This fact can be explained by the rather simple structure of the algorithm containing uncomplicated operations, namely, addition and shift (for quotient approximation), as well as shift and subtraction (for quotient refinement). Owing to CRT usage with fractions, the new algorithm does not include such operations as modular remainder calculation and number conversion into the mixed radix number system (MRNS) representation. The simulation of the algorithm on FPGA Kintex-7 has demonstrated a considerable reduction in hardware costs and an appreciable gain in speed as against the algorithms based on the CRT and MRNS representations.

Currently, this is the best hardware implementation of the general modular division. In comparison with the well-known algorithms, the suggested algorithm guarantees smaller hardware and time costs by a close connection between architectural calculations and hardware implementation. As a result, the computational complexity of modular division has been essentially decreased. The new algorithm is remarkable for easy implementation, thereby requiring fewer calculations than its well-known analogs.

A promising direction of further research is to find fast algorithms for several problem-causing operations in the RNS, namely, RNS-MRNS conversion and the optimal choice of RNS moduli within different ranges for specific applications. Each of the directions would promote the development of this field of computational mathematics owing to new RNS applications.

Author Contributions: Conceptualization, D.K., P.L., and N.C.; data curation, M.B. and I.L.; formal analysis, M.D.; funding acquisition, D.K.; investigation, A.L., A.N., M.V., and A.V.; methodology, P.L.; project administration, N.C.; resources, M.B. and I.L.; software, A.N., M.V., and A.V.; supervision, D.K. and P.L.; validation, M.D. and A.L.; visualization, A.V., M.V., and A.N.; writing—original draft preparation, N.C., P.L., and M.B.; writing—review and editing, D.K. and P.L. All authors have read and agreed to the published version of the manuscript.

Funding: This research was funded by the grant of the Russian Science Foundation (Project №19-19-00566).

Conflicts of Interest: The authors declare no conflicts of interest.

References

1. Omondi, A.; Premkumar, B. *Residue Number Systems: Theory and Implementation*; Imperial College Press: London, UK, 2007.
2. Szabo, N.S.; Tanaka, R.I. *Residue Arithmetic and Its Applications to Computer Technology*; McGraw-Hill: New York, NY, USA, 1967.
3. Molahosseini, A.S.; Sorouri, S.; Zarandi, A.A.E. Research challenges in next-generation residue number system architectures. In Proceedings of the 7th International Conference on Computer Science & Education (ICCSE), Melbourne, Australia, 14–17 July 2012; pp. 1658–1661. [CrossRef]
4. Mohan, P.V.A. *Residue Number Systems: Theory and Applications*; Birkhäuser: Basel, Switzerland, 2016.
5. Chervyakov, N.I.; Lyakhov, P.A.; Nagornov, N.N.; Kaplun, D.I.; Voznesenskiy, A.S.; Bogayevskiy, D.V. Implementation of Smoothing Image Filtering in the Residue Number System. In Proceedings of the 2019 8th Mediterranean Conference on Embedded Computing (MECO), Budva, Montenegro, 10–14 June 2019; pp. 1–4.
6. Younes, D.; Steffan, P. A comparative study on different moduli sets in residue number system. In Proceedings of the International Conference on Computer Systems and Industrial Informatics (ICCSII), Sharjah, UAE, 18–20 December 2012; pp. 1–6. [CrossRef]
7. Nakahara, H.; Nakanishi, H.; Iwai, K.; Sasao, T. An FFT circuit for a spectrometer of a radio telescope using the nested RNS including the constant division. *ACM SIGARCH Comput. Archit. News* **2017**, *4*, 44–49. [CrossRef]
8. Chren, W.A., Jr. A new residue number division algorithm. *Comput. Math. Appl.* **1990**, *19*, 13–29. [CrossRef]
9. Chiang, J.-S.; Lu, M. A general Division Algorithm for Residue Number Systems. In Proceedings of the 10th IEEE Symposium on Computer Arithmetic, Grenoble, France, 26–28 June 1991. [CrossRef]
10. Bigou, K.; Tisserand, A. Binary-ternary plus-minus modular inversion in rns. *IEEE Trans. Comput.* **2016**, *65*, 3495–3501. [CrossRef]
11. Lu, M.; Chiang, J.-S. A novel division algorithm for Residue Number Systems. *IEEE Trans. Comput.* **1992**, *41*, 1026–1032. [CrossRef]
12. Hung, C.Y.; Parhami, B. Fast RNS division algorithms for fixed divisors with application to RSA encryption. *Inf. Process. Lett.* **1994**, *51*, 163–169. [CrossRef]
13. Hung, C.Y.; Parhami, B. An Approximate Sign Detection Method for Residue Numbers and its Application to RNS Division. *Comput. Math. Appl.* **1994**, *27*, 23–35. [CrossRef]
14. Hiasat, A.A.; Abdel-Aty-Zohdy, H.S. Design and implementation of an RNS division algorithm. In Proceedings of the 13th IEEE Symposium on Computer Arithmetic, Asilomar, CA, USA, 6–9 July 1997; pp. 240–249. [CrossRef]
15. Bajard, J.-C.; Didier, L.-S.; Muller, J.-M. A new Euclidean division algorithm for residue number systems. *J. VLSI Signal Process. Syst. Signal Image Video Technol.* **1998**, *19*, 167–178. [CrossRef]
16. Hiasat, A.A.; Abdel-Aty-Zohdy, H.S. Semi-Custom VLSI Design and Implementation of a New Efficient RNS Division Algorithm. *Comput. J.* **1999**, *42*, 232–240. [CrossRef]
17. Gorodecky, D.; Villa, T. Efficient Implementation of Modular Division by Input Bit Splitting. In Proceedings of the 2019 IEEE 26th Symposium on Computer Arithmetic (ARITH), Kyoto, Japan, 10–12 June 2019; IEEE: Piscataway, NJ, USA, 2019; pp. 54–60.
18. Talahmeh, S.; Siy, P. Arithmetic division in RNS using Galois Field $GF(p)$. *Comput. Math. Appl.* **2000**, *39*, 227–238. [CrossRef]
19. Yang, Y.H.; Chang, C.C.; Chen, C.Y. A high-speed division algorithm in residue number system using parity-checking technique. *Int. J. Comput. Math.* **2004**, *81*, 775–780. [CrossRef]
20. Chang, C.-C.; Lai, Y.-P. A division algorithm for residue numbers. *Appl. Math. Comput.* **2006**, *172*, 368–378. [CrossRef]
21. Chang, C.-C.; Yang, J.-H. A division algorithm using bisection method in residue number system. *Int. J. Comput. Consum. Control (IJ3C)* **2013**, *2*, 59–66.

22. Chervyakov, N.; Lyakhov, P.; Babenko, M.; Nazarov, A.; Deryabin, M.; Lavrinenko, I.; Lavrinenko, A. A High-Speed Division Algorithm for Modular Numbers Based on the Chinese Remainder Theorem with Fractions and Its Hardware Implementation. *Electronics* **2019**, *8*, 261. [CrossRef]
23. Hitz, M.A.; Kaltofen, E. Integer division in residue number systems. *IEEE Trans. Comput.* **1995**, *44*, 983–989. [CrossRef]
24. Knuth, D. *The Art of Computer Programming, Volume 1: Fundamental Algorithms*, 3rd ed.; Addison-Wesley Professional: Boston, MA, USA, 1997.
25. Kogge, P.M.; Stone, H.S. A Parallel Algorithm for the Efficient Solution of a General Class of Recurrence Equations. *IEEE Trans. Comput.* **1973**, 786–793. [CrossRef]
26. Parhami, B. *Computer Arithmetic: Algorithms and Hardware Designs*; Oxford University Press: Oxford, UK, 2010; ISBN 9780195328486.

© 2020 by the authors. Licensee MDPI, Basel, Switzerland. This article is an open access article distributed under the terms and conditions of the Creative Commons Attribution (CC BY) license (http://creativecommons.org/licenses/by/4.0/).

MDPI
St. Alban-Anlage 66
4052 Basel
Switzerland
Tel. +41 61 683 77 34
Fax +41 61 302 89 18
www.mdpi.com

Applied Sciences Editorial Office
E-mail: applsci@mdpi.com
www.mdpi.com/journal/applsci

www.ingramcontent.com/pod-product-compliance
Lightning Source LLC
LaVergne TN
LVHW070151120526
838202LV00013BA/911